Robert J. LaGrange, Henry J. Jordan

Manhood

A Practical Medical Work on Nervous Debility and Physical Exhaustion

Robert J. LaGrange, Henry J. Jordan

Manhood
A Practical Medical Work on Nervous Debility and Physical Exhaustion

ISBN/EAN: 9783337373108

Printed in Europe, USA, Canada, Australia, Japan

Cover: Foto ©berggeist007 / pixelio.de

More available books at **www.hansebooks.com**

MANHOOD;

OR,

SECRETS REVEAL

A PRACTICAL MEDICAL WORK

ON

Nervous Debility and Physical Exh

TO WHICH IS ADDED

AN ESSAY ON MARRIA

WITH IMPORTANT CHAPTERS ON

DISORDERS OF THE REPRODUCTIVE

BEING A

SYNOPSIS OF LECTURES

DELIVERED AT THE MUSEUM OF ANATOMY, 70
PHILADELPHIA,

DRS. LA

(

Office, 16

We caution the Public and our Patients against all parties making use of our name and copying our lectures. As this has frequently occurred, we have taken precaution to enter all our new works in the office of the Librarian of Congress, at Washington.

DRS. LaGRANGE & JORDAN.

(LATE JORDAN & DAVIESON.)

TABLE OF CONTENTS.

(3)

CHAPTER VI.

CHAPTER VII.

CHAPTER VIII.

CHAPTER IX.

INTRODUCTION

To the One Hundred and Fifteenth Edition.

IN commencing our remarks on the highly important subject of MASTURBATION, and other diseases in connection with the reproductive organs, we cannot impress too strongly on our readers the Divine ordinance, "increase and multiply;" for by constantly bearing in mind the object for which we were sent into the world, it will at once be seen how sinful must be the practice of those who, by fatally anticipating the purposes of nature, are rendered incapable of procreation, and entail everlasting misery, shame, and ignominy on themselves here and hereafter. The seminal liquor, it must be observed, is the richest and most powerful of all the animal secretions; it is, in fact, the very essence of life; it is this fluid that strengthens our bodies, and by rendering the nervous system powerful, enables us to exercise our memory, imagination, and judgment for our worldly benefit and happiness. Now, if this important fluid be wasted in a manner which

(5)

nature never intended, and by which not only the fluid itself is lost, but the nervous energy is also impaired and exhausted by undue excitement, what results must follow? In the first place the nervous system of the masturbator becomes impaired; the brain, the heart, the lungs, become impoverished, and hence arise melancholy, impotency, a bewildered mind, nervousness, and general decay of the system. 'Tis then that the truth flashes across the mind, and the miserable victim of folly becomes aware of the extreme wretchedness of his situation, and that he is no longer a fit object for society; a complete imbecile, incapable of sexual intercourse; a man only in form, but not in substance, without the power of exercising his functions either of mind or body; the former participating in the disease, becomes morbidly affected, and distrust, fear, extreme sensitiveness, and frequently madness, ensue.

Here then arises a proof of the importance and necessity of the arrangement whereby some well-informed members of the medical profession may devote their exclusive attention to diseases arising from the undue excitement of the generative system, together with those incidental stages of acute disorder, which, if neglected,

terminate in the horribly wasting forms of constitutional disorganization.

To this part of the subject we have paid the
most anxious and untiring attention from a very
early period of our professional career. It is
one, in fact, that not a day passes in which we
are not consulted, either by professional visits,
or by correspondents in different parts of the
country, and we feel that we are not exceeding
the limits of truth, or transgressing the bounds
of professional etiquette, in asserting that our
mode of practice, suggested and improved by
long and multiplied experience, has been productive of the happiest and most successful
results in the treatment of sexual debility.
During our practice, also, we have too frequently
marked the great extent of constitutional disease, primarily springing from neglect or maltreatment of syphilitic diseases. Any medical
man who will make it his study, as it has been
ours, to investigate as far as possible in every
case, the original channel through which disease
or constitutional disorder first found its entry
into the system, will be astonished at the mass of
human suffering which may be traced to *venereal*
origin, although its primary symptoms may
have been for years apparently eradicated from

the frame. Nor do the sources of this misfortune lie very deep from observation. The malady generally commences its attack in early life, before experience has overcome the short-sighted heedlessness of youth, and taught it to look beyond the pains and pleasures of the passing moment. Delicacy or shame will not allow the sufferer to seek assistance until the poison has acquired a strength and virulence too alarming to be neglected ; the patient then, instead of applying to his usual professional friend, flies to some unskilful practitioner, who temporarily arrests the external symptoms, and discharges him as cured. Thus matters go on until the malady becomes constitutional; when the patient is at last compelled to place himself under the treatment of those who, at an early period, might have preserved his constitution untainted, and his body comparatively uninjured by the ravages of this insidious disease.

It is some years since the idea first occurred to us that a popular treatise, divested as much as possible of technical language, would be of much avail in counteracting the effects of the complaint, resulting from maltreatment or neglect, by directing attention to its early symptoms, and by demonstrating as clearly as possible the consequences resulting from neglect thereof—by the general contamination of the system which must necessarily follow.

Under this impression we have ventured to submit the following pages, and trust their utility may be acknowledged; we are desirous of explaining that they are not intended to supersede medical aid in any stage of the disorder, but that, on the contrary, we would impress upon our reader, if need be, the prudence of having immediate recourse to it in the earliest stages of the disease.

In the following pages we have as briefly as possible given a description of the organs of generation, and of their physiology, so that any one may readily understand their importance in the general economy ; and secondly we have brought under the notice of the reader the *causes, varieties, symptoms,* and *peculiarities* of these disorders of the generative organs, which either partially or totally obstruct them in the due discharge of those important functions which they are ordained to perform in the human system, and on the proper discharge of which not only rests the happiness of individuals in families, but also the welfare of empires ; for it is not to be disputed that on the degree of vigor and healthy action of those organs in the parents depends in a great measure the health of their offspring; daily experience presents to our notice painful and not unfrequently loathsome evidence of this fact.

Hence it behooves all persons, before entering into the married state, to inquire whether or not they are in such bodily health as may insure that the marriage bed shall not become a *hot-bed of disease*, whence naught but weak and puny offspring shall be produced—living evidences of folly and brutally selfish passion. How degraded and utterly lost to all finer feelings which alone can ennoble us, must that man be, who, knowing himself to be tainted by disease or so debilitated by early and guilty excesses that it is next to impossible that he should give life to any but tainted and doomed progeny—doomed in the mother's womb!—yet dares to offer his polluted and shattered frame at the pure shrine of female love.

To causes such as these may be traced much of the domestic unhappiness we daily see, and many of the serious diseases under which mankind suffers from one generation to another; therefore, reader, if you still be unmarried, let us beg of you to ponder well on these *truths—*

truths, which, if you neglect now, may at some future time *painfully and vividly be recalled in your own family.*

The fearfully abused powers of the human generative system require the most cautious treatment; our studies for many years past have been exclusively directed to the treatment of the debility and diseases resulting from self-pollution, venereal infection, loss of sexual power, and such complaints as arise from a disorganization of the reproductive powers, whether constitutional or acquired; and in conclusion we may observe that all who apply for advice or assistance may always depend on that inviolable secrecy, sympathy, and skilful attention which have always proved the basis of the most extensive practice in special diseases in the United States of America.

LaGRANGE & JORDAN, M. D.

1625 Filbert St., Philadelphia, Pa.

CHAPTER I.

ANATOMY AND PHYSIOLOGY OF THE GENERATIVE ORGANS.

In the present work, devoted to the consideration of special maladies, their causes, prevention and treatment, it will not be necessary to enter upon a descriptive account of the anatomy of every part of the human body; we shall therefore confine ourselves to that portion of the frame that is included in the subject-matter of our researches. We propose to treat solely of the diseases of the generative system, and their influence on the constitution and intellect; a brief description, therefore, of their anatomical and physiological relations will be all that is requisite. A brief but accurate description of those parts, and their uses in the human economy is necessary, in order that what follows may be fully understood.

The great importance of the organs of generation, and their preservation in a state of health and vigor, has been generally acknowledged; in fact, the due and proper performance of the special functions with which they are charged

has ever been considered essentially necessary
to the health and well-being of the economy
both physical and mental. They are of admirable construction, form and use ; and constitute
a striking evidence of the wonderful skill and
contrivance in the adaption of a special mechanism in the system for the performance of one of
its most important and essential functions—that
of the propagation of the species. Unequaled
in the delicacy of their texture, and the com-
parative minuteness of their structure, their
peculiar fitness for the functions assigned them
in the economy, when they are in a state of perfect integrity, excites the astonishment and
admiration alike of the anatomist and the phil-
osopher. Their very complexity, while it
renders them liable to many disorders, by any
of which their utility may be impaired, is wisely
rendered subservient to the important purpose
of separating and purifying the vivifying fluid.

The male organs of generation may be divided
into the *external* and *internal;* the former com-
prising the penis, urethra, scrotum and testicles;
and the latter or internal, consisting of a considerable portion of the urethra, the tubes
arising from the testicles, conveying the seed
to the seminal vesicles themselves, the prostate

and the verumontanum in the prostatic portion of the urethra, and indirectly all the organs engaged in the urinary affections, such as the kidneys, the ureters, and the urinary bladders.

The KIDNEYS, the organs solely engaged in the secretion of the urine, are glandular bodies of an oblong shape, seated on either side of the spine, upon and below the two last ribs, and behind the stomach and intestines, and are external to the peritoneum. The right kidney is below the liver when a man is in the erect position, and the left below the spleen; the right kidney is generally the lower and the larger. In shape this organ resembles the kidney-bean. The secretion of the kidneys is much influenced by the passions. We need only instance, in proof, the effects of fear on quadrupeds, infants, and even on men, in suddenly increasing the quantity of urine, and producing an insurmountable desire to void it. In patients laboring under some difficulty in passing urine, from the presence of one or more strictures, the mind referring to the complaint will often greatly increase the secretion of that fluid, and multiply the call to pass it from the body, and thus greatly add to the already existing irritation, and perhaps set up a new disease in the part.

The URETERS are long, hollow tubes, the continuation of the pelves of the kidney. There is one on each side of the body, and they pass downwards, and slightly inwards to the back and lower part of the bladder. Their use is to convey the urine from the kidney into the bladder.

The BLADDER is situated in the lowest part of the body, which is called the pelvis. It is of considerable size, and admits of distension to a degree that would hardly be credited, were it not a well-established fact. This organ in man lies directly on the bowels, but in woman the womb intervenes between it and the rectum. It is of an oval shape, and is the great receptacle of the urine. It has three coats, one of them being composed of muscular fibres, the construction of which causes the expulsion of the urine; it has, on that account, been called the *detrusor* or *expulsor urinæ.*

The PROSTATE GLAND, in shape and size, somewhat resembles a chestnut. It is situated below and behind the bladder, and above and in front of the rectum.

The URETHRA is a membranous canal, extending from the neck of the bladder to the end of the penis or yard; it is very vascular, and possesses a certain degree of elasticity. Its mem-branes are very thin, and almost transparent, and without fibres, so that in itself it does not possess the power of muscular contraction and relaxation. It is, however, provided with muscles, the action of which is to assist the expulsion of urine, and also of the seed during copulation.

The SCROTUM, or PURSE, is a bag of skin, divided about the middle by a septum, so as to form two cavities, in each of which a testicle is contained.

The TESTICLES, or organs which secrete the semen, are supplied with blood by long and tenuous vessels, which arise from the main arterial trunk, called the spermatic arteries ; the blood which they thus receive serves for the elimination and secretion of the seed,—a pro-cess which is effected by the peculiar action of the testicles, and which secreting power gives these organs a value and importance in the human frame not even second to that which attaches to those generally regarded by anatom-ists as the more noble, being those the de-struction or serious impairment of the functions of which may involve the loss of life. The value which men place on these organs (the testicles), or rather on the due performance of their functions, is rendered evident by the fact that suicide is not unfrequently caused by their supposed or real imperfection; and men on whom the operation of castration has been performed, in consequence of cancerous or other serious disease affecting the testicles, generally become moping and melancholy, and speedily perish. The same result occurs when, from a similar cause, the penis has been amputated.

Eunuchs, who have been castrated prior to the possession of those feelings which nature causes to spring up in man after the period of puberty, are of course not subject to the same degree of depression and wretchedness of mind and body as are those who are rendered impotent after having shared in the happiness and delight of matrimonial intercourse. Their disgust of life arises from witnessing comforts which others enjoy, from which they are forever debarred, but which they have had no means of fully appreciating. There is a marked difference in the external characteristics of a man and of a eunuch. The latter are rendered, by the degrading operation to which they have been subjected, more effeminate in personal appearance than are those who are in the full vigor

2

and enjoyment of manhood. The voice resembles that of children, the hair is thin and delicate, the limbs are small, the beard and whiskers do not grow, or at best are thin and scattered, and the mental faculties do not attain either vigor or penetration. Most of these changes and differences in the constitution not unfrequently follow the operation of castration, when performed during manhood, if it be complete—that is, if both testicles have been removed.

The testicles are generally two in number, one on each side of the serotum or purse; but cases have been published in which there has been only one testicle; and in others, again, there have been found three, four and even, although very rarely, five. The older writers, by whom some of these cases have been mentioned, considered the possessors of so unusual a number of testicles to be more than ordinarily salacious.

It occasionally happens that the testicles do not attain their full size and powers of secreting semen. This state has been termed *an arrest of development*—a phrase meaning that the organs at a certain period of life, prior to puberty, have ceased to grow; generally produced from an early indulgence in self-pollution. A case has been described of a gentleman who, when in his twenty-sixth year, had a penis and testicle which were not larger than those of a boy eight years old; and another of a man, thirty years old, in whom those organs presented a similar appearance. Such instances are not beyond the influence of medicine, unless, perhaps, when they occur in persons of idiots. Wasting or diminution in the size and power of the organs may occur at any age. The testicle is generally of the proper shape, although diminished in size, but feels soft, having lost its elasticity and firmness. It is pale in texture, and its blood-vessels appear to be less in size than in the healthy state. The secretion contained in the seminiferous tubes is entirely devoid of spermatic granules and spermatozoa, the nature and use of which will be mentioned in a short time. It sometimes happens that the organ undergoes what is called the fatty degeneration. The spermatic cord is usually affected by an extension of the disease; the nerves shrink, the blood-vessels are reduced in size and number, and the cremaster muscle disappears.

When disease of the organ is the cause of its atrophied condition, it becomes altered in shape, being uneven and irregular, and sometimes elongated, as well as diminished in size and weight. The proper glandular structure also seems to have nearly if not altogether disappeared.

Among the causes of this atrophy of the testicle, may be enumerated impeded circulation, local inflammation, whether arising from a special cause, or from the transfer of inflammation to the testicles. Excess in sexual intercourse and onanism are also efficient causes of an atrophied condition of these important organs. They will be alluded to more in detail hereafter. Their action is generally preceded by a low kind of local inflammation.

Injuries of the head, especially the back part, have not unfrequently been the cause of atrophy of these organs, and it has been known to occur without any apparent cause.

The fact that injuries of a severe nature affecting the back part of the head are sometimes followed by such a result, would tend to support the views of the phrenologists, who contend that the seat of sexual desire is in the cerebellum, which is there located, and between which and the organs of generation they say there is great sympathy. The brain, either in its entire, or in a particular part, undoubtedly exercises great influence on the desire for sexual intercourse. In fact, the influence of the mind on the organs of generation, and of the latter on the mind, is completely reciprocal.

So much similitude is there in the structure of the brain and of the testicle, as well as a most extraordinary sympathy between them, that experience in the course of a practice extending through a series of years, has demonstrated to us, that there are many cases where the human mind suffers under a species of derangement in consequence of diseases of the organs of generation. especially from *tabes dorsalis.*

The PENIS consists of the cavernous bodies (*corpore cavernosa*), and of the spongy body (*corpus spongiosum*), the latter terminating in the gland or glans. These are enveloped in a loose folding of common integuments.

The absorbents of the penis are very numerous, and terminate in the glands of the groin.

The spongy substance of the urethra, which forms the GLANS PENIS, is covered externally with an exceedingly thin membrane or cuticle, under which are placed the very sensitive nervous papillæ, which are the chief seat and cause of pleasure and pain in this part. We may now understand why many, in the venereal act, have not the GLANS distended, though the whole penis is, at the same time, turgid; because the GLANS belongs entirely to the cavernous body of the urethra, and if that body be paralytic or weakened from any preceding or existing cause—and this we have often known to proceed from irregular practices—so that the spongy body of the urethra cannot be distended, impotence will arise, which, if not perfectly understood, cannot be cured by any physician.

CHAPTER II.

OF SELF-POLLUTION OR ONANISM—THE CONCEALED CAUSE OF SEXUAL WEAKNESS, IMPOTENCY, LOCAL AND GENERAL DEBILITY, ETC.

Having in that branch of our professional duties to which we have more especially devoted our attention, witnessed the heart-rending effects resulting from the practices alluded to under the above title; and this, alas, in the majority of instances, from utter ignorance of the *sin* and *danger* thereby incurred, we feel that while writing a book on the subject of *Generative*

Infirmities, we should be neglecting our duty, did we not, to the best of our ability, show in glaring colors, the enormity of the sin, and deteriorating effects of "Onanism," that unnatural practice by which persons of either sex defile their bodies alone, in secret, whilst yielding to filthy imaginations they endeavor to imitate and procure to themselves those sensations which attend sexual intercourse; the habit ascribed by the poets to

"The solitary monk, recluse, obscene,"

and those who in the ardor of inconsiderate youth suffer themselves to be governed by passion rather than reason; whose sensual imaginations impel them to anticipate the ability of manhood, ere vigor has established its proper empire; demolishing the delicate groundwork of physical energy, soliciting an age of disgraceful imbecility, and bringing, ere middle life breaks on the summer of adolescence, all the sensible infirmities of senility; producing in its impetuous current, such an assemblage of morbid feelings, that life often becomes a wearisome burden, and its endurance beyond the power of reason to sustain.

This revolting and destructive vice, alluded

to in the 38th chapter, 9th and 10th verses of
Genesis, as the sin of Onan (hence its name), is
doubtless placed there for our warning, and as
an indication of our Creator's just abhorrence
of such unnatural sin.

This unfortunate delusion is usually first
communicated at schools, or public seminaries,
sometimes at the early age of nine or ten, before
the subject of it can be aware of its awful consequences, hence the absolute importance of a
virtuous education to restrain unruly passions
at this critical period, when youth begins to set
aside the authority of the parent—for precisely
as it is with Springtime, so it is with youth.
If the husbandman would have useful produce,
he must plough the soil and sow the seed, otherwise the Autumnal crop will be but weeds and
refuse; so is the period of youth (when the
conscience is tender, the heart susceptible, the
imagination vivid, and the cares of the world
somewhat distant), of vast importance, as the
favorable time for receiving the seeds of useful
knowledge, and right impressions, out of which
the bloom and fruitfulness of future character
may grow.   If the education be neglected now,
antagonistic principles, error, vanity and vice
will spring up as luxuriantly as the weeds of a

neglected field, furnishing a rich harvest of future regrets and sorrows; and with every delay in right cultivation, the problem becomes increasingly doubtful, whether human nature will ever grow anything but tares.

"Alas ! for those whose life and opening morn,
No type hath shown of nature's smiling Spring."

The indolent husbandman may retrieve his error by redoubled industry when the season comes round again; but youth goes, and never returns.

"For life, alas ! here knows no second Spring."

The principle of shame, and habit of self-denial, ought to be strongly impressed on the young mind before the arrival of this period of life. Dissolute companions, dalliance between the sexes, and those high-colored extravagant fictions which influence the imagination, and excite a state of morbid sensibility, ought to be cautiously guarded against. The passions awakened before their natural season, are the common destroyers of the youth of both sexes; and after they are fully established, too much attention cannot be paid to the choice of associates, the selection of recreations, etc. To keep the mind and body in a state of constant employment, and to observe temperance in diet and drink, are the most essential correctives, morally speaking, of the inexperience, enthusiasm, and impetuosity of young persons. These remarks are equally applicable to the correction as to the prevention of bad habits. This task is, however, by so much the more difficult.

As this disgusting habit commits the most unrestricted ravages upon youth, and inasmuch as it strikes at the very root of society, at the increase and propagation of the human race, by enervating and debilitating the springs of life, it will be obvious that no language can be sufficiently strong in its reprobation, for long experience teaches that of all the voluptuous and dangerous pleasures that strew the path of youth, none are so mischievous as that of Onanism, which unhappily offers two powerful inducements for its perpetration: first, it can be practiced in seclusion; and secondly, its effects on the health and personal appearance are not so immediately apparent, as, for instance, the paleness which succeeds a night of drunken and sleepless revelry. For a time, the solitary, vicious gratification may be concealed; the evil consequences are not known, and consequently not anticipated ; present excitement banishes the thought and fear of future suffering, but from the insidious manner in which this undermining process is going on, the truth will most assuredly one day present itself in awfully-distressing reality. The miserable sufferer is not sensible, it may be, for a long time, of the slow yet certain change that is passing over him ; the debility and paleness perceptible to others, have crept over him insidiously—no one part of the body feels weakened more than another ; as to the mind, however, the case is different ; a failure of memory being sometimes the earliest indication of mischief.

The evils resulting from self-pollution are twofold :—such is the mysterious nature of the union existing between mind and body, that any physically bad habit, while it undermines the bodily health, produces a corresponding depression upon the animal spirits ; the brain and nervous system becomes weakened and diseased, until one common ruin involves both alike in destruction. If self-pollution have unhappily gained the mastery over the young spirit, if it have become an admitted habit, the energies of the body, which ought to be generally directed to the purposes of nourishment

and growth, are employed in the reparation of
a criminal loss; and the purposes of natural
sustenance, as well as the support of the bodily
functions, are altogether superseded, or at least
imperfectly provided for.

An idea may be formed of the nature of this
loss, and of the sacred guard which health imposes
upon its due preservation, by observing the
consequences resulting from its unnecessary
and too frequent evacuation. It has been as-
serted by physiologists, that the loss of one
ounce of seminal fluid, by self-pollution or
nocturnal emissions, weakens the system more
completely than the abstraction of forty ounces
of blood; without lending ourselves to the accuracy of the extreme statement, it is sufficiently
clear that its due elaboration may be regarded
as of no small consequence to the system.
Hippocrates observed that "the seed of man
arose from all the humors of his body, and it is
the most valuable part of them." He says in
another place, "when a person loses his seed,
he loses his vital spirit," so that it is not astonishing its too frequent evacuation should enervate the body, which is thereby deprived of the
purest of its humors. Another remarks, "the
semen is kept in the seed vessels, until the man

makes proper use of it, or nocturnal emissions deprive him of it." Its importance, not merely for the direct end it was designed to fulfil in the process of generation, but for other purposes, is also indicated by the changes which take place in the animal economy at the age of puberty, when this fluid begins to be secreted; the voice and features change, the beard grows, the genitals become covered with hair, the whole body assumes a more rotund and manly appearance, the muscular system acquiring that firmness and solidity which chiefly mark the distinction between man and woman.

Loss of blood, if repeated, even though trivial in quantity, is a sure and readily-acknowledged index of corresponding failure of the vital powers; but the daily drain upon the nervous system from the undue loss of this most elaborate secretion, is still more rapidly destructive. The debility thereby produced, is greater than any other, inasmuch as important and extensive portions of the brain are concerned in its production. Physiology teaches us, that phosphorus enters largely into the composition of the brain and nerves, and as this substance also forms an essential element of the seminal fluid, the injury accruing to the system from any unnatural loss of this secretion is readily explained; and this fact alone affords a by no means insignificant indication as to the treatment of such cases. Who then can doubt the great importance of this fluid, or wonder at those evils its unnecessary evacuation is sure to entail?

Perverted indulgence in this horribly unnatural propensity undermines and poisons all enjoyment, inducing such misanthropic feelings, as absolutely to unfit the sufferer from all the ordinary business and enjoyments of life; so completely is the poor creature subdued by the wretched infatuation, that while conscious of the change that is taking place, he appears to have lost the power of self-control, or of making a proper effort to recover his position among his fellows. Torn by the contending passions of remorse and sensuality, his mind becomes the transcript of himself, moody, unhappy, ferocious, or miserable, distrustful, suspicious, gloomy, or childish; often a strange medley of them all, or presenting in the revolutions of a few brief hours, as many aspects of character as a fickle, uncertain, unmanly intellect can well appropriate.

"A withered frame !—a ruined mind !—
The wreck by passion left behind."

Sensibly alive to the impossibility of mixing in the ordinary enjoyments of life, and of deriving from sexual intercourse any of those thrilling delights which God, for the wisest purposes, has inseparably appended to that act, he becomes melancholy, dispirited, dejected; there passes over his mind a change which induces him to avoid all rational intercourse with his species ; the language of his actions is,

"Man delights not me, nor woman either."

He bids a gloomy farewell to the cheerful society and haunts of men ; the thousand anxieties and excitements of trade, politics, and commercial ambition, appear to his indolent imagination as either too great for his hopes, or foreign to his desires. Imbued with a moody misanthropy, the natural result of his own vices, he vents his splenetic complaints against the world at large, or peevishly declaims respecting the darker side of human feeling and character. Thus he becomes isolated, his mind vegetating on his own prurient and diseased fancies. Once, perhaps, there was the budding promise of future usefulness and activity ; now, how fearfully changed, the dupe of a lust alike horrible in imagination as well as in act. The blossoms of youth, perhaps the flower of manhood, the supremacy of the mind, all degraded, obliterated, gone!

Some continue the practice from feelings of despair. Conscious of its ruinous tendency, and desirous of resisting the unmanly habit, they have sought intercourse with women, but to their dismay, have found themselves power-less; and ashamed, vexed, dispirited, they forego any future attempt, lest they should again be subjected to the humiliation of failure. Abashed, the sufferer shrinks from the gaze of his fellows, fancying suspicion in the eye of every one who looks upon his sunken, haggard, pale, unmeaning, inexpressive face, his dull, lack-lustre eye, his thin and tremulous form—which all betray him to the practiced observer.

It is difficult to depict a more truly miser-able being than the slave of licentiousness. His imagination burning with filthy, unnatural glow; his bodily organs, taxed to the utmost, weary and jaded, refuse to obey the stimulus of that never-slumbering depravity, which goads his fancy in the darkness of night, in the dreams of his broken rest, and in the worse than dreamy abstractions of the cheerless day. Tormented with desires he can never gratify; shut out from those ˙enjoyments accorded only to virtuous moderation; and like Tantalus, thirst is consuming him, unmitigated by every attempt to force for a moment his mouth below the wave. The vulture retribution is preying upon his vitals, and furnishing him a striking fulfilment of the prophetic warning: " There is nothing done in secret that shall not be revealed," neither " hid " even from the recognition of mortals, that shall not ultimately be made, even to them, evident as the noonday.

Self-pollution entails upon its victims, marks as legible to the eye that can understand them, as the scars of small-pox. How much more perceptible to the eye of Him, by whom all actions are weighed—to Him, who knoweth the secrets of all hearts—from whose scrutinizing eye nothing can possibly escape! Can we produce a more fearful illustration of the stupefying effects of Onanism, than the fact, that the victim of this filthy abuse fears not in secrecy, though the eye of God is upon him, to do that, which if caught perpetrating, even by a child, or more especially a woman, he would redden with shame, and if possible, hide his head forever. Dreadful depravity! strange perversity! deliberately and secretly to deprive himself, by

3

a worse than suicidal madness, of the power of natural enjoyment—to entail misery upon himself in this world, and no hope of escape from condemnation in the next. I may here quote from the late Sir Astley Cooper, who stated in one of his lectures that "If one of these miserable cases could be depicted from the pulpit, as an illustration of the evil effects of a vicious and intemperate course of life, it would, I think, strike the mind with more terror, than all the preaching in the world. The irritable state of the patient leads to the destruction of life, and in this way, annually, great numbers perish. Undoubtedly, the list is considerably augmented from maltreatment, and the employment of injudicious remedies."

The first of the following cases is interesting, from the circumstances of an eruption on the face, to which the patient had been subject, being of undoubted syphilitic origin, derived hereditarily, though there are medical men who strenuously deny the possibility of this disease occurring for the first time at maturity—an undoubted fallacy, as we can prove. Many other hereditary diseases, such as gout, consumption, scrofula, cancer, etc., first show themselves at this period; and why should it be impossible in the case of syphilis?

The other cases, as embodying all the more prominent symptoms of the alarming infatuation we have endeavored to denounce, cannot be better given than in the very words of the writers themselves.

CASE No. 1477.

Pittsburgh.

GENTLEMEN:—I have read your Treatise, and wish that I had seen it earlier, and now hasten to lay before you my case; hoping you will advise me into the means, if means there be, by which I may regain my former health. Should your advice be a course which I can adopt, without too great restraint, and one affording no practical inconvenience to the duties of my public capacity, I will endeavor to fulfil it as honestly and frankly as I now make to you the following confession:

At a public school, and at the age of sixteen, I first contracted self-abuse. At eighteen, I left school, and for the space of a year gave myself up to a moderate use of this wretched habit. At nineteen, my health was so impaired, that a physician of high standing was called in, and my general debility then accounted for. (I, of course, keeping my own counsel.) Cod-liver oil, tonics, and a change of residence were recommended. I was accordingly sent to Florida. Here I became enamored with a very pretty girl, and then my habit became converted into a passion. That which I could not obtain in reality, I sought for in imagination and self-abuse. At twenty-one, I left for college. During my college life I still continued the practice of self-abuse, even up to this present time. So, from the age of sixteen to that of twenty-five, I have continued in a moderate abuse of myself. I have given it up for a month at a time, but so strong is the temptation, that even religious conviction can scarce overcome the will, and the force of conscience is entirely hushed by the dictates of a lustful imagination. I have now confided to you that which I never dared to breathe to any living creature, and even in this to you, I feel the acuteness of self-abasement and humiliation, which your kind counsel and treatment may, by the blessing of God, assist to remove.

I will now endeavor to give you a sketch of my personal appearance, age, occupation, and general symptoms. I am 27 years of age, unmarried. I stand nearly six feet; am dark, with a ruddy expression of countenance; very thin; gait stiff and wants elasticity and firmness; eyes generally weak, very black under the eyeballs, are also sometimes very hot and uncomfortable, and feel sore in their sockets. Hair is generally dry and for the most part thin and weak. My face, too, for the last five years, especially at Spring (just now) and Autumn, presents a very unpleasant appearance—red, scaly blotches of a syphilitic nature cover my chin and forehead. The former makes it very unpleasant for shaving. I showed it, about a month ago, to my "medical adviser" here (but did not tell him the suspected cause), he said it looked like secondary symptoms. I told him, however, what is the truth—that I had never contracted any venereal disease. He prescribed for me a decoction of Sarsaparilla, and I have been taking this with but little real benefit. I have for the most part a good appetite, and never suffer from indigestion; I am, however, very subject to cold feet, and the extremities, including my penis, are often very cold; of late, too, I have had sleepless nights, owing principally, I fancy, to want of circulation. I have, in fact, all the more common symptoms of general debility. My memory seems sometimes to be failing me, and after no very violent exercise, the blood immediately rushes to my face, and both my face and hands are often suffused with involuntary blushes. I am rather an irregular liver, fond of good ale, and good things in moderation; do not care much for wine or spirits, except whisky, a drink of which I take occasionally the last thing; I am a smoker, but this in moderation. I have plenty of good walking exercise, and live in a healthy part of the country. I am no lover of books, and am no advocate of sedentary habits; my bowels are generally very costive, with this, my spirits are depressed. Now let me add a description of that part immediately affected, i. e., my Penis. This, when in a pendulous state, is very small indeed, the end or glans is much swollen, and when an erection takes place, which is frequently the case, it is very hard and inflamed.

I have experienced such a thing as an involuntary emission caused by a dream, but such a thing is unusual; I always experience fearful erections in the morning when first I awake, and my eyes smart very much. My penis, when sitting, writing or reading, and in a depended state, frequently becomes hot, and unpleasant sensations of itching occur in the glans; when this is the case, I fear that external pressure would cause an emission. I care not, now, too, so much for women's society—I used formerly to desire it much. My urine, too, becomes affected, and has a strong smell. My principal ailments are want of circulation, frequent constipation, and .the unpleasant appearance of syphilis in my face. Now, Doctor, having given a faithful description of my symptoms, with a history of my wretched habit, I hasten to bring this unpleasant communication to an end. Enclosed is a post-office order for $5, the amount of your fee. And now, relying that you will assist me to break off that which is fast ruining my brightest prospects, and bringing me down to an early and ignominious grave. If you can guarantee, by your treatment, a speedy recovery, I shall ever consider you my best friend.

Ever faithfully yours,　A. B.

Drs. LaGrange & Jordan, 1625 Filbert St., Phila.

Case No. 1736.

Harvard, Mass.

Gentlemen :—I have been for years suffering from that species of disease to which you have

particularly directed your attention. The seeds
of it were implanted in my constitution when
quite a child, through the unprincipled conduct
of a domestic servant. When about fourteen, I
had an illness, which our physician traced to
this cause. I recovered to outward appearance;
but do not seem ever to have been thoroughly
cured. I now believe that I have been subject
to occasional emissions during the night ever
since that time; at least, I know I have been for
a considerable period.   *   *   *   A white,
fatty matter also collects between the glans
and the outer skin, and the glans is somewhat
inflamed. There is always more or less of an
eruption on my back; my face is also subject
to a similar eruption; my countenance is sallow
and unhealthy looking, and my face, as well as
the rest of my body, extremely thin. I feel
much weaker than I naturally ought. I often
experience a most uneasy sensation in my breast;
my hearing is impaired, and my voice has lost
its natural strength and clearness. My mental
faculties, too, are much clouded and weakened—
my memory, in particular, is affected. I am a
student—have studied pretty hard, and not very
unsuccessfully; but I have often felt an incapacity, and sometimes a disinclination for study,

which I knew were not natural, though I knew
not to what to attribute them till I read your
book. I have never been in the habit of using
tobacco, or drinking anything, save water.

I am engaged more than four hours a day in
teaching; during that time I scarcely sit down,
and have to talk a great deal. I am about the
middle height, rather slender, though naturally
strong and wiry. I am, I think, pretty easily
affected by medicine. I am afraid I have not
given you a sufficiently full and clear account
of my disease. Perhaps your experience may
enable you to propose a few questions, answers
to which might be of use to you in judging of
my case.

I await your answer with the greatest anxiety,
as I am afraid, from the disease having got so
confirmed in my system, that it will be scarcely
possible to eradicate it. But I beg, that you
will give me your candid opinion.

DRS. LaGRANGE & JORDAN, 1625 Filbert St., Phila.

CASE No. 1129.

Baltimore, Md.

GENTLEMEN:—I received your " Practical Ob
servations" from my friend, and its perusal
makes me hope that you will be able to afford

me such help as will bring me back again to the enjoyment of health and happiness.

I am twenty-nine years old. Since the age of about fifteen, I made myself guilty of the sin of self-pollution. Awakened to a sense of my position when being on the point of marrying, I had, in the summer of last year, recourse to Dr. ——, from whom I had a course of medicine which certainly had a favorable effect upon my system, as for a long time after I did not feel so thoroughly wretched as I had been before; and, hoping that the improvement noticed in my system, would finally prove a cure, I married in November last. To describe what I suffered since—how my sin stands before me night and day, in the most horrible forms, would be impossible; moreover I suppose you will be able to conceive what must be the sensations of a being, not entirely devoid of feeling, when he finds it impossible to respond to the loving embraces of a pure, innocent being, whose expectations of bliss and happiness in married life are most cruelly destroyed. A cruel doubt, inseparable from the state of mind of sufferers like me, and the impossibility of affording the expense of treatment, prevented me having recourse to medical assistance. Dr. ——'s recommendation has lighted a hope again in my heart. Oh! that I might, like him, have to thank you for a cure. Hoping this, I confidently place myself in your hands, and I beseech you to be kind enough to say if you think a cure in my case is possible. I trust you will charge as low as possible, considering that an innocent being, my wife, is dependent upon my exertions for her existence.

My symptoms are as follows: loss of memory; a continual feeling of languidness, tired with the slightest exertion; nervousness; when thinking about anything earnestly, my head turns quite dizzy. Nocturnal escapes have left me during two or three years but the penis is very small, and I am without feeling and desire.

I enclose your fee; and hoping to hear from you by return of post, I remain, etc.

DRS. LAGRANGE & JORDAN, 1625 Filbert St., Phila.

CASE No. 759.

Scranton, Pa.

GENTLEMEN:—I took some of your medicine some time ago, with considerable advantage, and should have availed myself of your advice before now, but could not do so comfortably on account of money affairs. About sixteen months

ago I left New York in a bad state of health;
my stomach, lungs, liver, etc., were thoroughly
out of order, having worked in a small ill-venti-
lated room, breathing impure air, making a very
hearty meal towards evening off beef, etc., but
scarcely eating anything next morning. However, I had saved money, had it deposited in
the bank, and what with doctor's bills, and being
out of health, causing me to neglect work, I
have exhausted my stock, so that if I do not
meet your demand immediately, I hope you will
have the kindness (though it is a great thing to
ask, being a total stranger) to give me a little
time to do so. I want my health improving, so
that I can use my fingers, and get on my legs
again. I enclose $5, and if you can indulge me
in the above request, I will pay you like a man;
feeling sorry that I did not hear of your skill
sooner than I did.

As to my symptoms—in the first place my
height is five feet nine inches, moderately fleshy;
age, twenty-nine (just), pale looking; occupation
a painter (but I would just say that I do not
feel much inconvenience at present in this respect, because I do not go to work before nine
o'clock, and can work as I feel disposed); habits
pretty regular; present symptoms are a little

nervous debility if in company; or if called to mental exercise, which is the case sometimes, I am seized with such an internal twitching, my lungs and stomach feel so racked, that I don't know where to put myself; when under this influence, my mouth becomes very dry, the breath raw and somewhat unpleasant, saliva thick, but when at ease mentally, it soon passes off, but I often feel a close morbid sensation in the lungs, sometimes pain, especially with some motions of the body or arms, which causes me to think my lungs not so strong as they have been. Drowsy, heavy sensations, which cause my eyes to be very dull, marked underneath with a little unnatural color. I am not troubled with a cough, but when I do cough it is very husky, etc. The Doctor told me my lungs were sound, but I do not feel them so healthy as I should like, if it can be remedied. I can take in a considerable quantity of air, and hold it a long time without inconvenience. My tongue is coated in the morning, tinged with yellow.

My parents are living, getting a good age, and we are as a family, healthy. When I was quite a lad, I contracted, from example, that horrid practice, continued to do so two or three years, but I became so disgusted with it, that I relinquished it; but I find that I have been troubled with the consequences ever since—want of manliness, fretting, heavy forebodings, etc. ; troubled with nocturnal emissions (but irregularly); sometimes pained a little in the left testicle ; a little fleshy substance just above, void urine well, etc. Hair grows pretty well, whiskers, etc. ; can walk ten or twelve miles without much fatigue. I am afraid you will think me tedious, but I feel anxious to give you every particular. I sleep well, eat well, but am troubled with internal morbid sensations ; I want to feel as I have felt, prior to twelve months ago, healthy ; manliness always deficient ; my teeth are in good condition, a thick matter accumulating around them.

Hoping to hear from you as soon as convenient, and that you will agree with my request,— I am yours, truly.

DRS. LAGRANGE & JORDAN, 1625 Filbert St., Phila.

Another patient writes—

CASE No. 1693.

Chicago, Ill.

DOCTORS :—I can scarcely express the shame and condemnation with which I have been almost overwhelmed for the last few weeks, since

I was made aware of the fearful consequences of that soul-destroying sin, " self-pollution," which I have been addicted to for the last seven years.

I am now twenty-four years of age, stand five feet nine inches in height, and weigh ten stone. My general health is good; my strength is also good, but my appetite fails me very often; by the time I have consumed two or three mouthfuls, I feel a desire to stop; my hunger is quite appeased; sometimes I feel quite melancholy; at other times I have a superabundance of animal spirits. I am not much subject to pain, but sometimes after I have committed the act, I have felt a slight pain in the left testicle, and at other times in the passage of the yard; at other times, I have felt a dull pain in the left side for two or three hours at once. I am subject to frequent and nightly emissions; in voiding urine, I have seen seminal fluid run away from me thin and unelaborated, especially when straining out the last few drops, and the end of the yard is almost constantly wet with fluid that escapes from me.

I have had much trouble and thought respecting one thing, and I do not know whether such a case ever came under your notice or not; it is this, the end of the yard still retains the foreskin or nut, the glans is not yet permanently denuded. Please inform me by return, whether such a thing will be troublesome, if I thought of being married. Do not be afraid of giving me your opinion ; let me know the worst. Would to God that you could cure me of this horrid malady. I was once subject to religious impressions, but now I may truly say with our Lord, " The last state of that man is seven times worse than the first." I shall know no bounds to my feelings, if you will kindly cure me. I may as well state my occupation, which I at first felt an objection to. * * I feel often very dull and heavy about the head as though from determination of blood ; and my superintendent often taxes me with being careless, when at the same time, it is through an impaired memory. I sometimes forget myself altogether, as being unconscious of persons being present.

All this is to be attributed to that horrid practice. I shall do my utmost to recommend others. I am under the impression that I can detect cases of this description, and will spend a portion of time in the work. I think I know, yes, I am sure of one. I do not think I can inform you of anything else. I enclose your fee, hoping to have a reply early, with the necessary instructions.

P. S.—As I am universally known in this part, please make use of the name, ———, in sending your packages, which I hope you will secure from observation, sending me word previous. Address ———. Letters may be addressed to me in full.

Such, and so sad, then, are the leading features of this soul- and body-destroying sin.

To different minds, however, it is necessary to produce varying argument and illustration: We speak at least to those who have not yet cast off all fear, all reverential regard for Holy Writ, and will, therefore, subjoin a few passages bearing evident reference to these crimes. St. Paul in the 5th chapter of the Epistle to the Ephesians, warns us, " Let no man deceive you with vain words ; for because of these things, cometh the wrath of God upon the children of disobedience." " Keep thyself pure," says the same apostle; and again, " Unto the pure all things are pure ; but unto them that are defiled, nothing is pure ; but even their mind and conscience is defiled."—1 Tim., v, 22; St. Paul's

4

first Epistle to the Thessalonians, chap. v. "And the things which he did displeased the Lord; wherefore He slew him."—Gen. xxxviii. "For this is the will of God, even your sanctification, that ye should abstain from fornication. That every one of you should know how to possess his vessel in sanctification and honor. Not in the lust of concupiscence, even as the Gentiles." Epistle to the Romans, chap. xiii, "Not in chambering and wantonness." "Walk not," says he, "as other Gentiles walk, who being past feeling, have given themselves over to lasciviousness, to work all uncleanness with greediness."—Ephesians, iv. 1 Cor. vi, "Know ye not that your bodies are the members of Christ? shall I then take the members of Christ, and make them the members of an harlot? God forbid." 1 Corinthians, iii, "Know ye not that your body is the temple of the Holy Ghost, which is in you, which ye have of God, and ye are not your own." And in the same to the Corinthians, chap. iii, "If any man defile the temple of God, him shall God destroy, for the temple of God is holy, which temple ye are." Doubtless many of these passages refer primarily to excesses with women, but they all point clearly enough, not so much to the act, as

to the filthy and polluted state of mind and
heart, out of which originate all the varied
forms of Sensuality, not excluding Onanism
itself, and which, if present in any of their disgusting characters, inevitably react upon the
mind, rendering it still more corrupt. Even
the Pagan world of ancient Rome, immersed
in sensuality almost beyond parallel, if we may
credit Martial, held the practice of self-pollution
in worse than abhorred contempt.

"Hoc nihil esse putes? Selus est mihi, crede; sed ingens
Quantum vix animo, concipis ipse tuo."

"We, however, have a more sure word of
prophecy, whereunto, if we take heed, we do
well." This ought to be a powerful induce-
ment to restrain us, not only from the grosser
forms of sensuality, but the more terrific practice of self-pollution; for God, as the creator of
this curiously contrived frame of ours, hath put
upon it a measure of His own glory, the shadow
of divinity. Our bodies are declared to be the
fitting residence of Himself; in His power and
presence He formed, and by that presence, sus-
tains us in being; wherefore our bodies, perish-
able and mortal though they be, yet partaking
of this honor, it is not less our duty than our

highest privilege, to retain them in honor. If the temples of heathens were not suffered to be profaned or polluted, how much more ought the thoughtless voluptuary to respect himself, if not for his own sake, yet as the curiously-constructed handiwork of that Being who will surely avenge Himself by pain and agony, inflicted upon the violator of His own laws. We are told in the unerring oracles of truth, "Ye are not your own, for ye are bought with a price, wherefore glorify God in your bodies, and in your spirit, which are His." Paul, speaking of heathens, observes that, "being given up to uncleanness, they dishonored their own bodies," and, in another place, "It is the will of God that we abstain from uncleanness." Indeed, the passages of Scripture, relating to the pollutions of the flesh, are almost numberless. Let individuals examine the Sacred Writings for themselves, and they will perceive that no other crime is so many times named as Uncleanness; and surely no person can be more accurately designated as guilty of this sin than he who is addicted to self-pollution.

If we reflect ever so superficially on the ordinary law of God's moral government and providence; if we think of the natural end and design of marriage in all countries, and the way in which it is evidently intended our species should be propagated, reason itself would instruct us that to destroy that end must be very offensive to our Creator, as well as a great crime against society. All form of sensual excess alike tend to destroy sexual power and the production of weakly, puny progeny, who are likely to bring no honor or usefulness to the State. If this be true of excess in general, it holds infinitely true of self-pollution, justifying the assertion that the crime is in itself monstrous, unnatural, filthy, odious, as its consequences are certain and ruinous, for it destroys conjugal affection, perverts natural inclination, and tends to extinguish the hopes of posterity.

The effects of this practice upon the body are not less remarkable than the strange debility which clouds the mind. And, be it observed, there is no act which so soon becomes habitual. In the first employment of tobacco and spirits, time is usually required to render them agreeable; but the first essay of self-pollution is ushered in with a new, wild and intoxicating delight. Its very secrecy aids the infatuation. The stream once crossed, the Rubicon once passed, all may be done effect-

ually that is evil for time and eternity. To re-
trace that step—to efface it as a blot from
memory and conscience is impossible; and so
often that inward monitor becomes seared,
deadened, hardened, till its feeble voice, from
oft-repeated criminality, becomes drowned in
the mad and urgently loud calls of unnatural
passion; and thus it is that the mind now depraved, becomes not the reasoning governor,
but the goad, the stimulus to acts, which sooner
or later, will abolish and destroy every vestige
of intellect. The uncleanness having obtained
the mastery over the heart, it pursues its victim
with lustful conception at all times, and in all
places, upon the most serious occasions, and in
the very acts of religion.

As the nervous system suffers, the brain becomes the subject of disease; melancholy, indifference, disgust, misanthropy, pass through
their various grades into madness, and the start-
ling truth must not be concealed, that self-pollution is a frequent cause of insanity. The late
Dr. Armstrong observed, "The solitary vice of
Onanism produces affections of the head;" and
in his published lectures is detailed the case of
"a youth of 17, who at the age of 10 was sent
to a school where he became the subject of this

vice ; and from a fine, active and clever boy, he
became an idiot. His eyes became prominent,
his pupils largely dilated; he had pains in his
head and down the course of his spine, loss of
memory, a silly, unmeaning expression of countenance and a tottering gait." The same writer
states, " I think I should know a person in the
street who has addicted himself to this vice, by
merely walking behind him, from his peculiar
gait."

As an illustration of the value and power of
observation, the above may serve usefully to
alarm some poor youth, who foolishly imagines
that his secret and soul-destroying infatuation
is known only to himself.

That these oft-repeated acts should really
tend to insanity, if we had not ample evidence
of the fact, it would not be unphilosophical to
suppose, for the mind, continuously directed to
this one morbid idea, becomes debilitated from
the perpetual recurrence of the same train of
thought ; and such is the sympathy of the generative organs, that the physical and moral
sensibilities are there directed to one common
focus, and that which ought to be only a casual
excitement, becomes exchanged for a permanent
morbid irritability. The purest and most ar-

dent love produces physical debility, evidently
because of the intense occupancy of the mind.
How much more debilitating must be impure
associations, connected as they are with vicious
practices? No wonder that madness and suicide
follow.

It is on record, that a gentleman in apparent
possession of every requisite to make life happy,
was found with a pistol (the instrument of his
death) clenched in his hand; none could account for the rash act, and doubtless, but for
his own revelation, it would have passed away
as the mere "temporary insanity" of the
newspapers. Upon a piece of paper in his own
handwriting were discovered the words, "I am
impotent and unfit to live." Scarcely a day
passes that deaths by suicide are not recorded,
and the laws of this country, as well as the
verdicts of a coroner's jury, generally place a
merciful construction upon an act which long
experience of such cases teaches us would be
better explained as arising from the sense of
impotence by self-pollution. The mental agonies
of such a one are almost insupportable. What
bodily pain can equal the agony of the soul,
aggravated as it must be by the consciousness
that to its own base sensualism he owes his

forlorn and miserable condition?—a being on whom the eye of beauty beams not with fond and pure affection—an outcast even from the paid embraces of a mercenary wanton.

Insanity, then, may be regarded as the sad and not unfrequent termination of these cases; and it must be obvious that the same causes which tend to break down the energies of the general system, will unquestionably tend to madness, and this because of the excitability of the nervous system, and the absolutely essential preservation of its powers. The more I see of such cases, the more I am convinced that the weakness of intellect, and even insanity, which such vice engenders, is often the sole and direct cause of suicide. On this subject Dr. Armstrong remarks, " I have met with many individuals, who have had, they say, a predisposition to self-destruction; and I find this especially the case where there is united disorder of the stomach, liver, bowels and head, which leads to madness."

We know that all efforts which require the exertion of continued imaginative power tend to madness. Poets, sculptors, painters, the writers of fiction, are all equally susceptible; and this from the exclusive exercise of the faculty of imagination.

"Great wits to madness nearly are allied,
And thin partitions do their bounds divide."

As to the moping, melancholy lover, his woes
are rational and natural, and find an easy termination ; but the solitary sensualist is the victim
of the worst, most unbridled and tyrannical lust
that imagination can embody ; every fair and
virtuous countenance that is new to him inspires
him with some filthy idea, excites and goads
him in secrecy to fresh excesses ; in silence he
conjures up before his diseased fancy, some
absent object ; the nervous system sinks under
the rapid unnatural whirl ; denied repose, urged
on to forbidden effort, tortured by the continual
recurrence of one single polluted idea ; reason
resigns her tottering throne, and staring madness, or it may be, muttering, moping idiocy,
usurps her vacant authority.

How fallen beneath the true nobility of man,
is the wretched wreck of humanity, whose excesses have reduced him to this contemptible
condition.   Once in the hilarity of youth he
rejoiced in the command of every faculty ; now
a senseless, yet animated mass of helplessness,
exciting the commiseration of those who know
not the cause of his ruin ; and visited with the
bitter scorn of those who, spite of his attempts

at concealment, read his degradation in his every feature. Whither may he fly from the plague that is within him, the evil that haunts him alike in darkness and by day? The quiet, refined enjoyments of literature, once his delight, now pall upon his morbid taste; if he read at all, nothing but the more licentious productions of our older dramatists, or the lewd effusions of the reign of Charles the Second, prove sufficiently stimulating ; or it may be these are exchanged for the mawkish sentimentality, the prurient voluptuousness, or concealed obscenity of a low circulating library of trashy novels. Forced to contemplate the gloomy spectre, the shadow of his former self, it is merciful indeed that loss of memory, in some faint degree, procures for him moments of repose from that murderous racking thought which can dwell alone upon images the most revolting! To such a one what misery arises from the accidental perception of domestic enjoyments ; he sees a fond father hug to his bosom his first-born, and cover its little laughing face with kisses. But for him—let fancy complete the picture.

It deserves to be attentively borne in mind that there is no vice which so certainly disturbs the natural circulation of the blood in the brain,

as the baneful habit of self-pollution, and inde-
pendently of the drain upon the system by
the frequent emissions of its most important
secretion, thereby occasioning great debility,
there is another and collateral cause of weakness
arising out of too frequent and unnatural erections, which have a tendency to generate a
species of atony, a palsied and enfeebled state
of the male organ, rendering it absolutely useless for coition. Hence, it frequently occurs
that on attempting intercourse, however strong
the previous excitement, the individual finds
himself suddenly impotent and powerless, not
so much from fear, anxiety or transport, but
doubtless from the absolute weariness and flaccidity of the penis, which, from mental causes,
had for many hours previously maintained a
state of throbbing erection. How much more
strongly must these remarks apply to the unnatural state of masturbators, who are liable
to an imperfect and temporary erection, even
from the utterance of a casual word. Such
frequent excitement brings in its train, not local
palsy alone (generative debility of the worst
kind), but general weakness of the severest
character.

Consumptive diseases are now so prevalent

as to justify every inquiry relative to their origin. There are, doubtless, some affections of the lungs closely resembling tubercular consumption, which are called into existence by the habit of self-pollution. And where the scrofulous tendency exists, we have no hesitation in affirming that this practice tends to its development. Sensual excess, as it is the most certainly fatal, so it is a most fertile cause of those diseases of the chest, which threaten so many in early life. Nothing can, therefore, be more irrational than the hope or expectation that the treatment of such diseases will be successful where, as in too many instances, the concealed cause (viz., self-pollution) has been overlooked and unconfessed.

Every person has his or her weak point, not merely in mind, but in body; but many persons from accidental causes (of which this vice is one', call into activity the seeds of disease, which would otherwise have lain dormant; and, as the result of this the early marks of some diseases of the chest may be noticed; breathlessness on the slightest exertion, irregular sleep, difficulty in falling asleep at night; together with these symptoms, there is drowsiness after rising; languor, lassitude, and other

signs of debility; fever is often perceptible,
chiefly in the evening. There is a loss of appetite, stomach is weak and irregular, and as the
food becomes less perfectly digested, a manifest
wasting of the muscular system follows. Paleness of the countenance; a tumid belly, with a
swollen condition of the legs; irregularity of
the bowels; frequent changes in the character
and appearance of the evacuations, are the pre-
monitory symptoms of consumptive diseases.
Presently, cough becomes troublesome; at first,
noticeable upon going to bed or rising; but
afterwards through the day. As the disease
advances, expectoration, first of a mucous, then
of a purulent character, ensues; and, before
death, purging and profuse night sweats leave
no doubt as to the real nature of the disease.
Early in the history of these cases, after fever
has begun to flush the check with a central redness, the urine becomes highly colored, and
deposits freely an earthy sediment. In many
instances, the stomach seems unimpaired; the
tongue, however, becomes smooth and unnaturally red; the voice is altered, and the eye assumes an unnatural brightness.

Epileptic and convulsive diseases are freely
produced, excited and called into action by ex-

cesses. The natural intercourse of the sexes is bounded by the natural capability, but this by none, hence there is excitement without power. Apoplexy occurs not unfrequently from this engorged and irregular condition of the blood vessels of the head, whether arising from Onanism, or mere venereal excess; the latter paroxysm terminates itself; the former, on the contrary, may be goaded on to madness; and if the vessels of the brain be not ruptured, it is that the most dreadful and exhausting debility remains behind.

Too great a dissipation of the animal spirits doubtless weakens the stomach, destroying the appetite; and as due nourishment no longer takes place, the motions of the heart and arteries become dull and languid. If the animal spirits and the seminal liquor be not identical, as the ancients supposed, yet this latter fluid like the blood, is undoubtedly alive, and vested with a vitality more intense, if possible, than that of the blood itself; so that the loss of semen produces the same results as the loss of blood, but in a greater degree. The celebrated Hoffmann relates that he has witnessed the most dreadful accidents result from an undue dissipation of the semen. "After frequent nocturnal pollutions," says he, "not only the powers are lost, the body falls away, and the face turns pale, but moreover the memory fails, a cold sensation seizes all the limbs, the sight is clouded, and the voice becomes hoarse; all the body languishes by degrees, disturbing dreams prevent sleep administering any relief, and pains ensue of the keenest description." He elsewhere observes that he had seen several examples of people, even after the body had attained its full vigor, who had not only brought on redness and acute pain in the eyes, but also such weakness of vision, that they could neither read nor write as formerly. It is recorded that "a young man, at the age of fifteen, having practiced masturbation till he was three and twenty, was afflicted with such a weakness in his head and eyes, that the latter were frequently seized with violent spasms. When he attempted to read, a kind of stupor occurred; the pupils became dilated and his eyes were exquisitely painful; the lids were heavy and full of tears, and matter gathering in the corners, which were painful. Though he ate with pleasure, he was soon reduced to a mere skeleton; and as soon as he had done eating, he was in a kind of intoxication."

The great Boerhaave has delineated these disorders with his usual strength and precision. He observes, " The loss of too much semen occasions lassitude, and renders exercise difficult; it causes convulsions, emaciations, and pain in the membranes of the brain; it deadens the senses, and particularly the sight, gives rise to a dorsal consumption, indolence and various other disorders which are connected with these. I have seen a patient," he says, " whose disorder began by lassitude, and a weakness in all parts of the body, particularly towards the loins; it was attended with an involuntary motion of the tendons, periodical spasms and bodily decay, insomuch as to destroy the whole corporeal frame ; he felt a pain in the membranes of the brain, a pain which patients call a dry burning heat which incessantly burns internally the most noble parts." Again, " I have also seen a young man affected with dorsal consumption. His person was very agreeable; he became so deformed before his death, that the fleshy substance which appears above the spinal processes of the loins was entirely wasted. The brain itself appeared to be consumed ; such patients, in fact, become insensible. They become so rigid, that I never perceived so great

5

a want of motion in the body, produced by any other cause."

Let not the intensely prurient, yet seemingly modest victim of secret pollution, lay the flattering unction to his soul that from the eye of his fellow-mortals he can conceal his unmanly practice. It is written upon his forehead—the physiognomy—that faithful mirror of the soul and body, gives clear indication of the internal disorder. The complexion and plumpness which jointly confer a youthful look, and which is the sole substitute of beauty; for without this, even beauty produces no other effect than cold admiration; this complexion and plumpness are the first things that disappear; a leanness succeeds, the skin becomes rough, often of leaden tinge; the eyes lose their brilliancy, and by their languor express that of the whole frame; the lips lose their vermilion hue, the teeth their whiteness, the hair falls off, and it is no uncommon thing for the whole body to become bent and distorted.

The observations of the ancients are in accordance with those of modern writers. Hippocrates has described the ills occasioned by self-abuse, under the title of Tabes Dorsalis. "This disorder arises from the spinal marrow,

and those who are given to unnatural enjoyments are afflicted with it. They have no fever,
and though they eat well, they fall away and
become consumptive. They feel as if a sting
or stitch descended from the head along the
spinal marrow. Every time they go to stool,
or have occasion to make water, they shed a
great quantity of the seminal liquor; they are
incapable of procreation; and they frequently
dream of the act of coition. Walking, particularly in rugged paths, puts them out of breath,
and weakens them, occasioning a ·heaviness in
the head, and noise in the ears, which are succeeded by a violent fever (lypiria), that terminates their days." And further, "that it is caused
by the wasting of the marrow of the back bone
in an unnatural way;" meaning doubtlessly, the
sperm, or seminal liquor. "The patient," says
he, "is free from fever, yet feels a kind of burning heat on some internal part; sometimes eats
and digests well; and if you ask him with
respect to his state, he tells you he feels a cold
running stream from the superior" or upper
"part of his body into the spine of his back,
and when he discharges his urine or his excre-
ment, there is sometimes an evacuation of
liquid semen. He is generally short-breathed.

feels languid after rising in the morning, with weakness about his loins, especially after much exercise, and his sleep does not afford him the wished-for refreshment. An intermitting dimness of the sight sometimes attacks him, his memory fails, and his spirits become dejected. This man," he continues, " will be incapable of propagating his species, unless the healing art afford him relief." He further observes, "that when this distemper continues for a length of time, it assumes various appearances in the constitution, and makes other stages under different characters; if not rightly understood it may end in an atrophy or nervous consumption; or perhaps in phthisis, or consumption of the lungs, where the healing art but too often proves unavailing." There can be nothing more dreadful than the picture which Aetius has left us of the evils produced by too great a discharge of the semen. " Young people have the appearance of old age. They become pale, effeminate, benumbed, lazy, base, stupid, and even imbecile; their bodies become bent, their legs are no longer able to carry them. They have an utter distaste for everything, are totally incapacitated, and many become paralytic. The stomach is disordered, all the body weakened; paleness, bodily decay, and emaciations succeed, and the eyes sink into the head." In another place he includes amorous pleasures among the causes which produce the palsy.

Galen has observed, that "the same causes occasion disorders of the brain and nerves, and destroy the powers of the genital system;" and in another passage he relates the case of a man who, not being quite cured of a violent disorder, died the same night he attempted the act of coition.

Pliny informs us of two individuals who died while attempting to force the powers of nature in the sexual act. These testimonies of the ancient physicians are confirmed by innumerable modern writers.

Sanctorius observes, " This weakens the stomach, destroys digestion, obstructs that insensible perspiration, the irregularity of which produces the most fatal consequences; occasions the liver and reins (kidneys) to be overheated, gives a disposition for the formation of stone in the bladder, diminishes the natural heat and usually occasions the loss of, or at least weakens, the sight.

Lommius, in his commentaries on Celsus supports the testimony of his author thus:

"Frequent emissions of the seed, relax, dry up, weaken, enervate, and produce a crowd of evils. Apoplexy, lethargy, epilepsy, fainting, loss of sight, tremors, palsy, spasms, and every species of the most racking gout."

Tulpius writes—"The spinal marrow does not only waste, but the body and mind both equally languish, and the man perishes a miserable victim."

"Nothing," says De Louvain, "weakens the stomach, and abridges life so soon."

Blanchard has been an eye-witness to gonorrhœa produced by self-pollution, to consumption and dropsy flowing from the same source; and in the memoirs of curious naturalists we find mention made of a person losing his sight from similar causes, thus demonstrating the close connection and sympathy existing between the testicles and the rest of the body, particularly the eyes.

Numerous other writers might be quoted, and instances adduced, in evidence of the injurious effects of this practice, and we cannot fail to perceive the accuracy of Galen in remarking, "This humor is nothing less than the most subtle of all others; it has veins and nerves which convey it from all parts of the body to the genitals. When a person loses his seed, he loses at the same time the vital spirit, so that it is not astonishing that too frequent coition should enervate, because the body is thereby deprived of the purest of its humors."

And Pythagoras observes, that it is "the flower of the blood;" a figurative expression, but one which accurately designates its noble office; in a word, it appears that the semen is the most important and elaborate of the animal fluids, the dissipation whereof leaves the other humors weak and vapid.

We need no better proof of the importance of the seminal fluid, than the results arising from its too frequent and unnatural discharge, as arranged by the eloquent Hoffmann.

I. " Involuntary nocturnal emissions ; pains in the back and often in the head ; weakness of the memory and sight ; and a mucous discharge from the urethra, especially after straining at a discharge of the excrements ; an aching rolling, and a pendulous condition of the testicles. The testicles being the secretory organs of the genital fluid, are furnished with arteries, veins, lymphatics and nerves, like other glands, and are suspended by the cremaster muscle. When, therefore, from any cause the texture is weakened, a pain will be felt, and this is increased by their hanging down, caused by that general relaxation of fibres that effects the whole body, and the cremaster muscle in particular.

II. "All intellectual faculties are weakened; loss of memory ensues, the ideas are clouded; the patients sometimes fall into slight madness; they have an incessant irksome uneasiness, continual anguish, and so keen a remorse of conscience that they frequently shed tears. They are subject to vertigo; all their senses, but particularly their sight and hearing, are weakened; their sleep, if they can obtain any, is disturbed with frightful dreams.

III. "The powers of their body decay; the growth of such as abandon themselves to these abominable practices, before it is accomplished, is greatly perverted. Some cannot sleep at all, others are in a perpetual state of drowsiness. They are affected with hypochondriac or hysterical complaints, and are overcome with the accidents that accompany these grievous disorders—melancholy, sighing, tears, palpitation, suffocations and faintings. Some emit a calcareous saliva; coughs, slow fevers, and consumptions, are chastisements which others meet with in their own crimes.

IV. " The most acute pains form another subject of patients' complaints; some are thus affected in their heads, others in their breasts, stomach and intestines, others have external rheumatic pains; aching numbness in all parts of the body, when they are slightly pressed.

V. " Pimples do not only appear on the face (this is one of the most common symptoms), but even suppurating blisters upon the nose, chest and thighs; and painful itchings in the same parts. One patient complained even of fleshy excrescences upon his forehead. All these symptoms show to what an impure condition the blood is reduced.

VI. " The organs of generation also participate in that misery of which they are the primary cause. Many are incapable of erection; others discharge their seminal liquor upon the slightest titillation and the most feeble erection, or the efforts they make when at stool. Many are affected with a constant discharge, resembling foetid matter or mucus, which entirely destroys their powers. Others are tormented with painful priapisms, dysuria, stranguaries, heat of urine, and difficulty in passing it, which greatly torments them. Some have painful tumors upon their testicles, penis, bladder, and spermatic cord. In a word, either the impracticability of coition, or a deprivation of the genital liquor, renders every one imbecile and impotent who has for any length of time given way to this crime.

VII. " The functions of the intestines are sometimes quite disordered ; and some patients complain of stubborn constipations ; and others of the hemorrhoids (piles), or of the running of a fœtid matter from the fundament and diarrhœa."

Another writer thus describes the evils arising from masturbation : " The muscles of the youth become soft, he is indolent, his body becomes decrepid, his gait sluggish, and he is scarcely able to support himself. Digestion becomes enfeebled, and the breath offensive, the intestines inactive, the excrements hardened in the lower bowel, producing additional irritation of the seminal vessels in its vicinity. The circulation being no longer free, the youth sighs often, the complexion is livid, and the skin assumes an unhealthy hue. The angles of the mouth become sharp and lengthened, the nose reddened at its extremity ; the sunken, glazed eyes deprived of all brilliancy, and encircled by a livid, lilac, or purple zone, are cast down ; no look remains of

gayety—the very aspect is forbidding and criminal. General sensibility becomes excessive,
producing tears without a cause; perception is
weakened, and memory almost destroyed. Dis-
traction, or absence of mind, renders the judgment unfit for any operation. The imagination
gives birth only to phantasies and fears, without
grounds; the slightest allusion to the dominat-
ing passion produces motion of the muscles of
the face, irregular spasms, the flush of shame,
or a state of despair. The desires become ca-
pricious, or envy rankles in the mind; perhaps
there ensues a total disgust. The wretched
being finishes by shunning the face of men, and
dreading the observation of women. His character is entirely corrupted, or his mind stupefied.
Involuntary loss of the reproductive liquid takes
place irregularly during the hours of sleep, as
well as on any attempt to evacuate the bowels,
and there ensues a total exhaustion." In such
patients there is an extreme susceptibility to
external impressions, as (*e.g.*) those arising from
variations of temperature, moistness or dryness
of the atmosphere. The mind becomes conscious
of those trifling changes of weather, which men
of business, happy in their vocation, regard not.
The irritable condition of the mucous membranes

of the genital organs is propagated, by sympathy
to the eyelids and nostrils; these become weak,
red, and watery, and are susceptible of colds
from very slight causes—so, at night, the glare
of a light becomes offensive and painful to the
eyelids; and even in the day a more frequent
winking and discharge of tears occur, if there
is the slightest wind; besides irregular flying
pains, often mistaken for common rheumatic
affections.

One of the effects of self-pollution is the actual reduction in bulk of the male organ; and
what is worse, its power of erection becomes
correspondingly destroyed. If we reflect upon
the difference between masturbation and the
natural act, we shall not wonder at this. Such
a one, if the seed vessels are not sufficiently
distended with the fluid that excites erection,
is able by unnatural friction, to produce a momentary discharge when nature refuses the necessary firmness for coition. In this way a host
of evils is engendered. The testicles are called
upon suddenly and violently to secrete a thin,
effete, unprolific fluid, and the nerves of the
penis are rendered susceptible of an agreeable
titillation without that natural adjunct, firm
erection; hence when the votary of self-pollu-

tion attempts intercourse, there is an absence of the requisite rigidity to effect penetration ; or if he partially succeed, a premature emission takes place, almost or altogether unattended by the slightest sensation of pleasure.

Among the minor evils, we must not omit those eruptive diseases, chiefly of the face, frequently observable among young persons, and often assignable to improper habits.

From time immemorial, the popular belief has been that the undue loss of semen from masturbation, or sexual excess, has a tendency to destroy the growth of the hair, and to produce baldness. According to Hudibras,

> " Want of virility is averred
> To be the cause of want of beard."

Nor is this opinion without some foundation in truth. Its presence in profusion is usually an index of sexual power, and when, from excess, that energy falters, nature, as if for the purpose of economizing her scanty resources, casts off the comparatively unimportant appendage ; the hair becomes white from defective nutrition, or in middle life the head assumes the baldness, though not the venerable dignity of age. The absence of hair upon the cheeks and chin is frequently associated with solitary practices; a beardless chin and an effeminate voice are the aversion of females, as well as the object of their ridicule, and they are generally pretty good judges that way.

All the ills occasioned by excesses with women, more quickly follow in youth the abominable practice of self-pollution, and which it would be difficult to paint in colors so glaring as they merit; a practice to which youth devote themselves, without knowing the enormity of the crime, and all the ills that are its physical consequences. The soul is sensible of all bodily disorders, but particularly of those which arise from this cause. The most profound melancholy, indifference, even aversion for all pleasures; the impossibility of sharing in the conversation of company, wherein they are always absent; the thought of their own unhappiness; the despair which arises from considering themselves the architects of their own misery, and the necessity of renouncing the felicities of marriage, are the fluctuating ideas which compel these miserable objects to sequester themselves from the world; and happy are those who do not put the finishing hand to their existence.

That masturbation should be so much more deadly and destructive than even excessive enjoyment with women, is explained by the fact that the latter has its limits of capability, whereas the former has none.

Scirrhosity, or a hard, enlarged, incipient, cancerous condition of the prostate gland, is not unfrequent among men in advanced life, particularly those who have imprudently produced excitement of the parts by long toying with women, or by Onanism. Such, then, and so sad, are the consequences of those unnatural sensual enjoyments altogether the reverse of that transporting emotion incidental to the caresses of a pure and virtuous affection, which, in some measure counterbalances the luxurious fatigue consequent upon rational and temperate indulgence. To this delicate susceptibility the miserable victim of solitary vice is a stranger. The warm and passionate kiss, the unutterable and thrilling embrace that only lovers can feel, live but in his diseased fancy; for it cannot be questioned that nature allots more joy to those gratifications, procured in her proper channels, than in those which are repugnant to our natural organization. The joy which only the heart can appreciate (and which must be carefully distinguished from the voluptuousness purely sensual, which even a prostitute may inspire), animates the circulation, aids digestion, accelerates all the functions, restores strength, and supports it. This it is that gives to marriage that sacred home-felt sweetness which love inspires and God looks down upon approvingly. The debauchee, scoffing at that he can never know, affects to despise marriage, because, owing to the degradation of his soul, there is a purity, and consequently an intensity in such interchange of affection, he can never realize.

Unfortunately, masturbation is not confined to the male sex. Deeply must it be deplored that the practice of vicious indulgence has found its way to the chamber of unmarried girls ; and the great, great responsibility of parents and guardians, in reference to the hidden character of those to whom is entrusted, not merely the physical, but the moral and intellectual companionship of artless youth, can hardly be adequately estimated. Surprising artfulness and obstinacy are employed by young people in maintaining secrecy respecting crimes of this description. Let the proud father of the clever girl, whose early spirit he is the first to appreciate, watch closely the associations she may form, even with those of her own sex; and most especially the books that she reads when no eye is near.

The same influences that entail early decrepitude upon boys, are in operation among girls to an extent which can only be believed by those whose peculiar province it is to conduct such inquiries; and the same causes which occasion impotency in the one sex, are productive of barrenness in the other, besides laying the foundation of those tedious discharges, and other serious ailments which undermine the constitution and embitter the existence of so many of our fair sisters.

There are results common to both sexes; but among females addicted to solitary sexual pollution, one of the first consequences is the formation of an irritable, nervous, bilious, hysterical constitution; and owing to the morbid accumulation of excitability in the genitals, there is a copious discharge of mucous fluid, generally only an augmentation of the natural secretion, but which, when excessive, is termed, in the closet-vocabulary of women, " the whites."

To bring these observations to a close: it is impossible to regard the gloomy consequences of self-pollution otherwise than as a direct retri-

6

butive infliction from the hand of that Being, whose marvelous handiwork, in the construction of the human frame, led one of old to exclaim, "I am fearfully and wonderfully made." Let, therefore, self-respect and adoration of the Great Artificer restrain the thoughtless, or it may be that driveling, helpless, premature dotage may (like the mark set upon Cain) place them in position past all hope or remedy.

CHAPTER III.

SPERMATORRHŒA, IMPOTENCY, STERILITY, ETC.

The genital fluid being retained (for subsequent excretion) in the vesicular receptacles, and through the removal by absorption of its watery particles becoming more concentrated, acrimonious, and powerful, gives rise spotaneously, and at certain intervals, according to the health and constitutional stamina of the individual, to the natural desire for sexual intercourse. These preliminary remarks are essential to a rational comprehension of that morbid habit of parts from which originates the disease or infirmity, called in professional parlance, "Spermatorrhœa."

The act of evacuation becomes pernicious in proportion as it is unnecessary, and not demanded by the wants of the living system; hence it is, that the youth who is devoted to unnatural excess, inflicts such retributive vengeance upon himself; he has the power of exciting these organs, when, from previous evacuation, they contain nothing sufficiently stimulating to cause erection, and maintain requisite condition for congress; in this way he is able to perpetrate a great amount of mischief, and the irritated vessels pour out a thin mucous discharge, possessed of no vivifying power. From this very irritation, the vesiculæ seminales or receptacles above alluded to, become incapacitated for retaining the semen as it is conveyed thither from the testicles; and should this miserable fraction of humanity marry, a mere thin gleety drain is all he can furnish to impregnate, it may be, a warm, vigorous and healthily organized woman. The solitary creatures who have reduced themselves to this deplorable state of helplessness, have drained the system, by past excess, of that which the seminal vessels are not, naturally, too irritable to retain, but which, if now secreted of a healthy character, is expelled without effort, or the excitement inseparable from the generative act. Here, then, is indicated one among the many forms of seminal weakness. Irritability is its leading feature, and among the more lamentable diseases to which the human body is liable, there are none requiring a greater degree of patience, discrimination and tact on the part of the medical man, and fortitude on the part of the patient, than this condition of the generative apparatus.

A stain upon the linen, and a strange feeling of weariness, are the first things that arrest attention; with this there is often an absence of that natural erection, which generally occurs in the morning, after first waking from the supine position, so that though each repetition of the weakness is noticeable in the early stages of this vicious perversity of parts, after a time, as the fluid escapes unconsciously, its effects upon constitutional power are the only signs, yet they are fearful tokens of its debilitating presence.

In many instances the sleep is not broken, and it is sometimes difficult to ascertain how often the evacuation occurs; the consequences of the loss are, however, sufficiently evident, demanding the most prompt and energetic measures to avert the impending mischief.

Nocturnal emissions, though more frequently

attributable to the practice of self-pollution, and sexual excesses, may arise from diseased testicle, or prostate gland. When complicated with the latter, the discharge of semen is mixed with the natural secretion of the prostate, and the combined fluids stain the linen of a dingy yellowish hue, closely resembling that resulting from the discharge in gonorrhœa (common clap) and the gleety discharge which accompanies its chronic stages. Lodgments of hardened feculent matter in the large intestines, sometimes act as a mechanical irritant, and give rise to diurnal as well as nocturnal evacuations of this highly important fluid, and hence the complaint may be said to possess three different stages, showing themselves—

1st. By nocturnal emissions.

2d. By diurnal losses (called Spermatorrhœa).

3d. By impotency, or loss of manhood, partial or complete.

All the three stages are accompanied by nervousness, more or less; impaired nutrition; lassitude; weakness of the limbs and back; indisposition and incapacity for study or labor; dulness of apprehension, deficient power of attention, loss of memory, aversion to society, love of solitude, timidity, self-distrust, dizziness, headache, pains in the sides, back and limbs, affection of the eyes, pimples on the face, and, in extreme cases, even by idiocy and insanity in their most intractable forms. It is generally attended with loose and dangling testicles, coldness of the glans penis, unfrequent weak and languid erections of comparatively short duration. The progress from one stage to the other, is often so gradual, as to be scarcely perceptible; the seminal losses being unsuspected, and, as it were, so hidden from view, that only a well-experienced eye is able to trace them. In the first place, the testicles have, through the practice of Onanism, or other excesses, acquired such a morbid sensitiveness, that on the slightest local irritation, they put in action their secretive powers, thus constituting an infirmity which might not inaptly be termed a consumption of those glands, which, if unimpaired by chronic disease, or pernicious practices, are fitted to supply and expend, at proper and chaste intervals, that important secretion during the most active years of a long life. But if manhood be anticipated; if the secretion and expulsion of semen be premature or unnaturally frequent, if it be excited artificially, nocturnal emissions may occur, as the consequence of the practice, but seminal or generative debility most certainly will.

These nocturnal emissions then cannot so easily be overlooked, but diurnal emissions or the second stage, frequently altogether escape the attention of the patient, till he finds himself in the third stage, and partially or wholly incompetent for the performance of sexual intercourse.

The emissions, at first attended by erections and pleasurable sensations during sleep, in time begin to occur without either erection or sensation, and ultimately take place during the day, whenever the bowels are moved, or the urine is passed, or during the slightest excitement in female society. In extreme cases, there is an almost constant discharge, or oozing of semen, and a complete absence of the power of retention. Persons who were formerly annoyed by frequent nocturnal emissions, and are not so much troubled in this way, are often perplexed as to the cause of their continued nervousness, incapacity for study or business, depression of spirits, aversion to society, etc., little suspecting, that at the very time they were congratulating themselves on their release from their former annoyance, they were enduring one of still more importance, and that the loss of semen, which was formerly only occasional, in nocturnal emissions, is now almost constant, it being carried off with the urine, at each evacuation of the bladder. This is easily explained. The semen passes from the testes (along the ducts, called the vasa differentia, opening into the urethra); these ducts, in a healthy state possess sufficient powers of retention, but when weakened from abuse, they become relaxed, and, as it were, dilated, allowing the semen to escape involuntarily on the slightest excitement. In this irritable condition, they are liable to be acted upon by the urine as it passes over their sensitive outlets; and this irritation extending to the bladder, the urine is in consequence voided frequently, and nearly always mixed with semen. It is only since the important aid to diagnosis rendered by the marvelous powers of the microscope, that this very important fact has been made out, prior to which, numbers have suffered all the consequences of debility, without a suspicion of the actual cause. It is no uncommon circumstance for patients to remark that their urine is thick and ropy, particularly the last few drops. This, in some instances, arises from inflammation of the bladder; but in most cases, this condition is caused by the semen it contains. Many married men, even of chaste and temperate habits are thus affected, without even suspecting it, simply because they are unacquainted with the existence of such a hidden source of weakness; and on their attention being drawn to the subject, they can but regret the want of previous information as to their increasing and apparently uncontrollable debility. In married people this frequently arises where the bounds of moderation have been overstepped; and there is no doubt that this hitherto undetected drain upon the system, has been the cause of incalculable misery to thousands.

Many men, about the age of forty, who have lived freely, are not unfrequently greatly altered in their power of sexual intercourse. They may, indeed, in general health, be stout and hearty, and for several years not very sensible of diminution in this respect; but the frequency of their inclination for such duties gradually becomes less, a symptom at all times portentous of approaching impotence; for the inclination ceasing, the power soon follows. In others about the same time of life, the physical power ceases first, and the inclination continuing often for years, their amusements become but the

merest pantomimes of amorous indulgence!
Such individuals otherwise in tolerable health,
are recoverable. From actual observation, we
have generally found that these different conditions owe their origin, and even their immediate
existence, to nearly the same causes, and may,
in almost every instance, be removed by nearly
similar treatment.

The term impotence is applied to this inability
or incapacity on the part of the male to the
performance of the sexual act; and it is requisite that for all practical purposes, we do not
confound this state of the generative system
with sterility, since a man who is sterile, or a
woman who is absolutely barren, may yet be
perfectly capable of the act of coition; in other
words, sterility is an inability for propagation:
impotence, an inability for copulation, whether
occurring in either sex, whether natural or acquired, whether resulting from disease or malformation.

The desire to perpetuate our species is one of
the most intense and irresistible passions with
which we are endowed; it is a part of our existence, it is a consequence, a feeling as natural as
hunger and thirst. The time at which this desire commences is Puberty, the most critical

period in the life of man; at this epoch our frames are perfected, the secretions necessary for the formation and growth of our bodies have, to a great extent, performed their office, and a total change, mental as well as corporeal, takes place. The boy suddenly throws off the puerile character, his whole appearance undergoes a change—his countenance is illuminated with intellect and decision, his voice assumes a rough and manly tone, his cheeks and lips become shaded with delicate down, the precursor of the distinctive beard; his limbs become firm, his step erect and vigorous, and he no longer delights in those occupations and amusements which before afforded him gratification.

In the female the characteristic changes are equally marked, and, if possible, the body undergoes still greater alterations; the system becomes fully developed—the bust enlarged, the eyes sparkle with vividness and expression, indicative of soul and feeling; the periodical indisposition peculiar to her sex commences, girlish playfulness is exchanged for that graceful bashfulness and retiring modesty which are so pleasing in girls of this age, her mind is occupied with ideas pure, but strange and absorbing; in a word she is a woman, "fairest of creation, last and best of all God's works."

Puberty is the most critical period of life. The mind rushes into a new world—new thoughts new feelings engage the attention, and the foundation of future character and happiness is now in the balance. The body participates in the change; the buds of inherent or acquired disease are now matured or crushed, and the prospect of continued health and strength will be influenced in a considerable degree by the conduct of life at this era. How important, then, that the opening mind and expanding reason of youth should comprehend the pinnacle upon which it is now poised! Surrounded by all the temptations and inducements to err, which on every side allure the inexperience and indecision of boyhood, it cannot occasion surprise that

"Some begin life too soon –like sailors thrown
  Upon a shore where common things look strange."

Dear is the price hereafter to be paid for this precocity; imprudence or excess may be indulged in with apparent impunity, while strength and youth have the power to neutralize the immediate effects of folly; but when these are exhausted, and disease turns the balance, rapid is the onslaught, and, it may happen, decisive the victory.

As this work is intended to be practical we have no wish, nor intention, to gratify prurient curiosity with any curious detail, but at once plunge in *medias res*, and ask—Is there a divinity, law or medical student, who does not aspire to a mitre, the woolsack, or a chair? Is there a mercantile drudge who does not aspire to be at the head of a firm? These are partial hopes, to be obtained only by few. Is there one man— however exalted, however humble—who does not look forward to the possession of a home, a wife and children, as the goal of his endeavors, his toils, his cares? This is but a general desire, within the reach of all. Home! wife! children! are the talismanic words which have guided men to the noblest actions—to the greatest efforts of genius and exertion. All worldly happiness is centered in these blessings, for what can excel the domestic comforts of his own fireside?

"———— ———— All who joy would win
Must share it. Happiness was born a twin."

We need not enlarge on this topic. Every man's heart must acknowledge its truth; and were we to quote from the immortal Milton, even to the veriest tyro of the Poet's Corner of a country newspaper, we could not improve the impression which must be spontaneous in the breasts of all. But decidedly, although Lust is not of necessity the constant attendant on Love; still "Love brooks not a degraded throne;" and that, as well to fulfil the Divine command, "Increase and multiply," as to fulfil the very purpose of our existence by continuing

> "—— The vigorous race
> Of undiseased mankind,——"

it is imperative to preserve in the perfection of their power those organs, and the functions of those organs upon which our obedience to this command depends.

Among the causes which impair and destroy this necessary attribute of man, it has been shown that self-pollution unfortunately takes the first place; and although the actual effects of this debasing practice may not have evinced themselves by any sensible derangement of the general health, still the source of the sexual debility may be no less certain. Impotence from this cause arises from actual want of power; from premature or tardy emission; from the impoverished state of the vivifying fluid; from disproportion in size of the organ; from the absence of desire; and from the universal weakness which affects the whole system.

Excess in venery is another cause of impotence, and is also a most fruitful source of some most serious maladies not peculiar alone to the male sex. The diseases which are consequent upon this excess are those which are dependent on the nervous system; namely, Paralysis, Apoplexy, Epilepsy, Diseases of the Organs of Vision, particularly Nyctalopia, and those spectres or dark spots which float before the eye, called by physicians Muscæ Volitantes. Diseases of the heart are greatly aggravated by excess in venereal indulgence. Whether the baneful habit of self-pollution, or excess with women have engendered this disturbed balance, the effect is the same. Sexual power is sure to be destroyed in the end. On attempting intercourse with women, the semen is too quickly discharged; nocturnal emissions sometimes occur, almost too frequently to be recounted; even the sight of a fascinating creature is sufficient to arouse the dormant irritability, and diurnal losses also take place. With some, erection is seldom, or exceedingly weak, and desire is more or less extinct. The mind partakes of the prevailing imbecility; the man, sensibly alive to his calamity, is reduced to the condition of a monomaniacal nervous hypochondriac, the seminal fluid dribbling away without erections, unconsciously, and not as the natural ejaculatory effort of the muscles appropriated for its convulsive discharge. No man is justified in entering upon the responsibilities of marriage, whose condition even approaches distantly to this. Before he enters into that most solemn engagement, it is his bounden duty, as he would avoid the most refined cruelty to an innocent yet affectionate woman, to ask his conscience well and truly whether there be any bar or impediment to that sacred union; and if suspicion be even delicately and tremblingly alive, let him wait until reassured of his partially lost powers, that he may with confidence lead his blushing bride to the matrimonial altar. If otherwise, the nuptial bed of the helpless unmanly creature, instead of creating a secret, yet intensely transporting delight, will be converted into a scene of blended mortification, disgust, disappointment and suppressed anger. Now it is that the fair girl laments, when tears are unavailing; now the cheated bride is made to feel herself the unhappy victim, the scapegoat of a long round of past sensuality—the sins of former years are made to tell upon her devoted head, if not in the communication of actual dis· ease, at least in the deprivation of those enjoy· ments which to a certain extent are essential to the happiness of wedded life. Anxious for off- spring, yet baffled from year to year in the feeble embrace of the man she has vowed to love and honor; life, health, and youth, fast wearing away, under a combination of circum- stances so painful, that language cannot ade- quately describe them.

The reproductive power may not be entirely destroyed by that state of generative debility engendered by nocturnal emissions, and yet very painful consequences of another character may unquestionably arise. A healthy female may become pregnant from the feeble yet ex- haustive efforts of a man whose constitutional power is seriously broken, yet it would be un- philosophical and unsupported by any analogy drawn from the history of the lower animals, to expect that this circumstance would not tell most detrimentally upon the offspring, which will assuredly bear enstamped upon it the same physical characteristics as the feebly vital fluids from which it originated. As illustrative of this,

7

it has been remarked, from the days of Aristotle, that illegitimate children are frequently endowed with great genius and valor. The circumstance has been ascribed to the impetuosity of both parents during their embraces. Hercules, Romulus, Alexander, Themistocles, Jugurthra, King Arthur, William the Conqueror, Homer, Demosthenes, and many others, are notable instances; and the most ancient families, in almost every kingdom, have sprung from the left-handed offspring of princes. The worthiest captains, best wits, greatest scholars, bravest spirits in English annals, have been base born. Carden, in his " Subtleties," gives a reason: " These are more noble and powerful in body and mind, chiefly from the vehemence of the sexual act that begat them." Probably their superior energy may be attributed to the strength of parental constitution, which is all for which we contend, the weak and delicate not being so likely to become the prey of unlawful passions.

Impotence is more frequently observed in our own sex than among women. Temporary impotence in man or woman, especially in the former, is often the result of mere apprehension, or of some diseased condition with which sexual intercourse is for a time incompatible; such, for instance, as nervous and malignant fevers; while, strange to relate, an opposite effect is sometimes produced by other diseases, such as gout and rheumatism, hemorrhoids, etc., and instances are on record, of other diseases producing such a change in the constitution, that an impotent man may find himself cured of his impotence on their cessation.

Impotence in the male may arise then from a wide diversity of conditions. Incapacity of erection, generally referable to self-pollution; impotence, arising from a want of power of retention in the seminal vessels induced by a morbid susceptibility of those vessels, brought on by a persistence in the same vicious practices; impotence, from inability of retention resulting from repletion of those vessels. Impotence from mental influence has also its appropriate management. Exclusive of this, the generative infirmity under consideration, though occasionally arising from simple disease, is ascribable in by far the greater majority of instances to the excess of sensualism, either with women, or more commonly still, from that vile delusive mentally annihilating excess, to which such frequent allusion has been made.

That long-continued debauchery, whether with women or by masturbation, is among the most common and prominent of the causes of impotence, is a fact admitted by all systematic writers. Mons. Pinel observes : " The impotence caused by the latter excess, reduces youth to the nullity of old age, and is too often incurable."

It is certain that where impotence arises from Onanism, its severity is far greater than when produced by excessive venery. The reason is, that the vital fluid, which should have improved the stamina of the system, has been lost without satisfaction; without that gratification of the mind which compensates in some measure the expenditure of vital energy. All authority favors the assertion that moderate legitimate indulgence with a beloved object, tends to produce a pleasing hilarity, lightness of heart, and aptitude for the ordinary business and enjoyments of life; and as the best things become by abuse or excess, the worst and most enervating, it is especially so in the present instance. The man who indulges to the utmost his licentious propensities, and taxes his failing powers to their last extent, seeking for variety as a new stimulus, may certainly find therein an excitement sufficient for the occasion, and may be able to effect and accomplish more frequent repetitions of the sexual act than the sober, quiet, married man, who, happy in his choice, is faithful to one woman. But the fact must not be withheld, that this excess among men of polluted minds is accomplished at the expense of a corresponding amount of physical energy; such feeling becomes the most dominant rampant lust, and no passion more strangely wears down its victim, strewing in prostrate wreck all the finer and more delicate emotions of the soul. It is a morbid craving which can never be appeased, and its end is not only the destruction of all mental quietude, but is the utter ruin of the body. Even when physical capability is expended, and premature decrepitude approaches, powerless desires still torment the victim of ungovernable lust, who can talk only of past enjoyments, but whose filthy conversation serves to inflame and seduce other and purer minds. The results of such efforts tell with tremendous power upon the greedy but hopelessly debilitated votary of pleasure. The lawful partner of a husband's bed silently affords only that gratification demanded by the sexual organs when fully charged with the seminal fluid, and impatient for relief. To a man so happily circumstanced the stimulus of variety is unsought, contemned, forbidden, not merely as contrary to all laws, divine and human, but as directly opposed to the maintenance of his animal organization in health, strength and usefulness. Here, then, the natural laws of his physical constitution harmonize most admirably with the higher sanctions of morality. The actual amount of enjoyment realized by the temperate is, in the long run, far greater; power is maintained until old age, and a vigorous offspring is engendered; while the hasty, violent, and forced gratifications of the sensualist, though intensely vivid for a brief space, are succeeded by that worst form of helplessness, insatiable desire appended to diseased and powerless organs.

And here we have a few words for those who, although strictly faithful to the marriage bed, are yet accustomed systematically to infringe the laws of chastity by indulging their animal propensities to excess. This also brings its punishment. Since the last edition of this work, a somewhat marked case of paralysis has occurred to us, which we will briefly relate. A gentleman from New Orleans had for some time lost (almost entirely) the use of both lower extremities ; not progressing very satisfactorily, a physician, whom he consulted, induced him to undertake a journey, with a view of placing himself under our care. He candidly admitted at his first interview, that having been exceed- ingly happy in his married life, it being his good fortune to possess a most amiable and affectionate wife, and being naturally of warm and ardent temperament, he had indulged his propensities to great excess ; to such a degree was he debilitated, that as well as the absence of the power, he then seldom ever experienced any sexual desire ; and this, in conjunction with his almost childish helplessness, debarring him from active exercise of any kind, so preyed upon his spirit, that he feared his mind would ultimately fail, and whilst writing he expressed an earnest desire to remain under our personal care. This gentleman remained in the city under our treatment for ten weeks, and then returned home, taking with him a further supply of our medicines. Some weeks afterwards we received from him the following communication, comment on which is unnecessary :

" I row, and swim, and ride, to the astonish- ment of all my neighbors. I also walk much firmer, enjoy my food as I have not done for months, and am already like another being. Thank God, I am improved in strength of body, and in tone of mind, beyond anything I expected in the time; I feel returning health coursing through my veins; only sometimes I ask myself if it is a dream. It is almost incredible to myself and friends, the benefit I have already derived from your treatment."

The draining of the seminal fluid, which occurs either from excessive venery, or from self-pollution, is not equally great in every instance. There are some individuals who are not rendered absolutely, but only partially, impotent.

With severe effort they accomplish the sexual act occasionally, though probably to the ill-disguised indifference, if not disgust, of the female. Others again, though unprolific, are tolerably competent, but at long intervals; their powers, though weakened, are not altogether destroyed, for with due care, and the really steady employment of judicious measures, this threatened evil may be averted.

Promptitude in these cases is of the first importance, while on the other hand, if valuable time be squandered now in contending against improper treatment, it is too probably lost forever. In just such cases as these it is that the cautious skill of science leaves blundering, bungling quackery far behind to pursue her blind injurious course. Here the great end and aim of treatment must be to excite without irritating; and to individuals menaced with this evil, I address the friendly warning, on no account to tamper and temporize with their infirmity in the idle hope that all will in time recover itself, for sad experience proves to the contrary.

Genital malformation—as amongst the proximate causes of Impotence—must not be passed without notice. The prepuce may be adherent to the glands, or so bound down by the frœnum, that without a trifling operation, copulation cannot take place. Phymosis and Paraphymosis may also render an attempt at coition excessively painful to the female and devoid of all pleasurable excitement on the part of him who labors under the infirmity. The penis may be curtailed of its fair proportions, or may have acquired such "lusty stature" as to offer another obstacle. Some patients have complained of serious inconvenience from an unnatural curve of the penis, not the result of disease.

Then there are disorders of the urinary or genital organs referable to inflammation or irritation of those structures; as, for instance, thickening of the bladder, disease of the testicles, wasting of the penis, stricture, with other affections of the organs. There is also a frequent affection of the prostate gland of a chronic cancerous nature, which forms a barrier to copulation. Cases in which the urethra terminated at the base of the penis, near the perineum, have come under our observation; these, though very distressing, are not irremediable. The testicle is sometimes malformed, and occasionally one or both are wanting. All cases of so-called Hermaphrodism are dependent upon malformation.

It would be possible to go on enumerating causes and effects in relation to this subject, almost without end; and although such cases may not in every instance be amenable to treatment, yet there is much room for ingenuity; and it has occurred not unfrequently, that means have been employed to very tangible purposes in cases previously regarded as hopeless. The advice, the consolation, the comfort to be afforded in all such cases are in themselves of no small moment, and as the "Medical Adviser" should be the friend, as well as the physician, of those who seek his aid, such investigations may not be found altogether so useless as at first might appear. We have been repeatedly consulted on account of the absence of offspring, by patients, who were impotent from one or other of the foregoing causes, and frequently has it been our happiness to render very essential aid towards the realization of this long, ardent, and very natural desire. Such cases call for no small degree of delicacy, tact, discrimination, caution and sympathy—and with these combined, much may often be accomplished by the humane and scientific practitioner.

Having thus noticed the chief of those causes which detract from the happiness and injure the health of man, we cannot conclude more appropriately than by quoting from Dr. Johnson's admirable tale, Rasselas: " Let us, therefore, stop, while to stop is in our power; let us live as men who are sometime to grow old, and to whom it will be the most dreadful of all evils to count their past years by follies, and to be reminded of their former luxuriance of health, only by the maladies which riot has produced."

Fortunately, the numerous records in our possession of cases which (though in some instances very serious) have been treated to a favorable termination, prove that the recovery of the powers of manhood is not (under judicious management) so utterly hopeless as might seem to be the case, trusting only to the observations of some medical writers on this subject.

Among women, impotence can only depend upon vicious conformation, mostly natural, but sometimes acquired, and the accidental result of dangerous lacerations or inflammatory adhesion, following child-birth; however, the causes of sterility or barrenness are numerous. One of the most frequent, being the malady called leucorrhœa, or "whites." Barrenness is also often caused by retention, irregularity, or profusion of the menstrual secretion, thereby giving rise to the diseases known by the names of chlorosis, or green sickness, amenorrhœa, or obstructed menstruation, dysmenorrhœa, or painful scanty menstruation, and menorrhagia, or excessive menstruation and flooding. It may also proceed from frigidity of temperament, or from aversion, reserve or indifference, which renders them insusceptible of anything more than mere passive submission, instead of appetency and pleasure; in short, it is a general irrespondence in the feelings of the female to those of the male, and

when accompanied by sterility, too often gives
rise to reproaches, dissensions and even disgust
—converting love to hatred, making the nuptial
couch a bed of thorns instead of roses, marriage
a curse rather than a blessing.

There are other causes of barrenness, over
which the female has no control; for example,
the womb itself may be unnaturally small; the
ovaries may be absent; or the fallopian tubes
may be closed. If these things co-exist, as the
consequence of the close and intense sympathy
existing between the female uterus and the
breasts, the latter are observed to be correspondingly small; and, contrarily, a free development of these, if it be not a mere fatty accumulation, is fair indication of a healthy natural
condition of the female reproductive system.

The analogy and mode of testing the physical and maternal capabilities, which many engaged in the breeding and rearing of cattle
have drawn from the development of the breasts,
is by no means unphilosophical or indelicate;
any causes, therefore, which tend to interfere
with these organs are important. A well-known
aged agriculturist observes on this subject, " I
am afraid that some of the defects of the French
women are to be found among the superior

classes, particularly in this country. The young women are generally much more flat-busted than they were sixty years ago. I now see them with different feelings, but I can observe forms with the same eyes, and several observant women have noticed the change. Look at the pictures of a century or a century and a half ago, and the bosoms of women there represented are not similar to those of modern times." How far secret indulgence in unnatural practices may have tended to this, is a question which the experience of those who are devoted to such inquiries can only decide.

We have endeavored, as succinctly as possible, to offer an intelligible portraiture of the interruptions to sexual health, and by explaining the causes in a simple, forcible, and perspicuous manner, to enable the reader to disentangle the apparently inextricable and confused maze of his own wandering and diseased fancies; to point to the concealed, and it may be unsuspected, cause of suffering, to the restoration of health, vigor, usefulness, activity, and joyous hilarity. Why do I suffer?—why, when all around me invites to enjoyment? why is it, that while every face wears a smile, existence is to me a weary blank—the world, its pleasures, cares, and duties, an irksome weariness? Are not these questions which even a cursory glance at the previous pages will enable the misguided to solve? Long experience of human nature, long acquaintance with some of its most painful infirmities, enable me to say it will be so.

CHAPTER IV.

THE TREATMENT OF SPERMATORRHŒA, MENTAL, MORAL, DIETETICAL, ETC.

In reference to which, I wish to impress the fact, that a most important object to be gained, depends entirely upon the patient, that is moral restraint, confirmed by the most determined resolution. The desires and passions must be controlled; allow them once to gain the mastery, and then farewell to health and peace. After perfect and continued abstinence from this pollution is insured, and not till then, we may commence with confidence the medical treatment; in doing this, we have two objects to attain; first, to renew and invigorate the general health, without stimulating or exciting the parts more particularly implicated; secondly, to remove local diseases and derangement, and to restore

the natural functions of all the organs. In
furtherance of the first indication, it is necessary that the patient banish from his mind all
melancholy, unnecessary forebodings and dis-
couraging doubt, as to the speedy and effectual
cure of his case; he must summon hope, and
occupy his thoughts as far as practicable, with
pleasing amusements; in furtherance of which,
cheerful society should be courted. Whatever
tends to excite the passions should be sacredly
eschewed; and every circumstance likely to in-
cite him to a repetition of his folly, should be
carefully guarded against. The standard works
on Zoology, History, Natural Philosophy, and
even those of the best novelists (forbidding all
of sensual character) may with advantage occupy much of his leisure.

The newspaper—that mighty organ of good
and evil, attracting the attention of all classes
of society, giving direction to human opinion,
and influencing a world of mind, should be read
in some intervening moments. In like manner
may be perused one or more of the numerous
admirable periodicals which grace our age. As
with the choice of reading, so it is · with the
choice of companions. One cannot be long in
the company of another, without being either

the better or the worse; if, therefore, you would improve by them, accept the counsel of the wisest of men: " Go from the presence of a foolish man, when thou findest not in him the lips of knowledge." Time is surely too valuable, and the space allotted to man too short, to fritter it away in the society of triflers; and above everything should the companionship of those be shunned, whose conversation is polluted by profaneness or licentiousness. Young men, especially, are very apt, when together, to indulge in loose conversation, little thinking that hardly anything so argues a degraded mind, a filthy taste, and a foul heart; for " out of a pure heart can come forth nothing but what is pure." Besides, such conversation is so unmeaning, so useless, so wanton, so vile, and so subversive of everything that is good, that it should never be countenanced by any who possess one spark of virtuous feeling; for—

———————— " When lust,
By unchaste looks, loose gestures, and foul talk,
But most by lewd and lavish acts of sin,
Lets in defilement to the inward parts,
The soul grows clotted by contagion,
Imbodies and imbrutes, till she quite lose
The divine property of her first being."

8

A youth on beginning the world, loses many pleasures and many safeguards too, when he shuts himself out, or is shut out, from the humanizing influences of family life. It should be the first anxiety of parents, if possible, to secure for their children at this most critical period, a residence in a well-regulated family circle, where they will be blessed with social comforts, and with the gentle presence of virtuous female society. From neglect of this simple precaution, many a man has had to date his first departure from the path of purity and his final overthrow in the vortex of dissipation.

The exercises of the mind, and those of the body, should be so regulated, that the one may serve as a recreation to the other; when the circumstances and opportunities of the patient permit, change of scene and air is highly desirable; moderate exercise on foot or in a carriage should be taken daily, while riding on horseback should be indulged in with caution.

It is greatly to be deplored that a more liberal feeling does not exist among those who employ young persons in business, than to exact fourteen or sixteen hours for labor or confinement, which must be not only injurious to health, but subversive of morality; and from my experience in diseases which are peculiarly solitary, I am of opinion that an immensity of evil may be traced to this over-taxation of time and labor.

It unfortunately happens, that sufferers from this class of diseases seldom feel so ill as to seek medical assistance, till, by continued protraction, they have greatly injured their constitution, as also the generative organs themselves.

It often happens, also, that through the diversified forms of disease resulting from self-pollution, the symptoms so closely resemble those of diseases of an entirely different nature, that the attention of the ordinary practitioner is not directed to the real disease, and mere symptoms engross his attention, while the organs actually suffering wholly escape notice. Hence the evil proceeds, and if its character be eventually suspected, it is at a period when the constitution has been so far invaded, that its condition demands our immediate and earnest attention. Unfortunately, neither the sedative powers of leaden girdles, the anti-spasmodic virtues of camphor bags, nor the cooling properties of nitre, have more than the imaginary effect in subduing this powerful passion, which requires the exertion of the rational faculties to give a proper direction to the laws of nature.

Now, what are the indications of treatment? Why, to remove the provoking causes that brought about the first estrangement, and to cultivate the mind into a higher notion of its own importance, as nothing is so apt to produce a relapse as a return to former practices; reason must resume her seat, and become a convert to chastity and honor. And here we would particularly remark, that the system generally adopted of employing the ordinary tonics to overcome seminal weakness is seldom productive of any permanent benefit. The seminal fluid is obtained principally from the most vital portions of the blood, and it is an excessive drain upon this essential secretion that thus fearfully weakens and frequently renders wholly inert the generative organs. It is by the employment of peculiar combinations possessing singularly nutritious, warm, invigorating virtues, that this important deficiency can be supplied, and the consequent depreciation in the character of both solids and fluids, effectually overcome. So to act upon the seminal vessels, as to impart tone and strength, without producing irritation; or temporarily exciting the

generative power to renovate the impaired
functions, by the exhibition of such remedies
as shall remove the proximate causes of debility
and disease, and thus permanently restore the
lost energies of the system,—is the treatment
which alone can be successful; and which the
experience afforded by several years' careful observation, in a practice of no inconsiderable extent, proves to be better adapted, and more
uniformly applicable to every variety and complication of debility, particularly that resulting
from the excesses to which we have made such
especial allusion in the preceding pages, than
any medicine or remedy for such ailments, with
which the long course of experience has made
us acquainted.

The more powerful of the ordinary aphrodi-
siacs are in the same ratio injurious to the general system, inasmuch that like most strong
stimulants, they are inevitably succeeded by a
corresponding depression of the vital energies,
which ultimately resolves itself into confirmed
and irretrievable impotency, and hence it is, and
very properly, that they are now so seldom employed. Our medicines, on the contrary, gradually impart strength to the organs, through the
constitution, which they first invigorate; and

though gentle in their action, are all the more permanent in their effects; rendering real service without the possibility of harm to the system. To individuals of either sex, even in the more confirmed cases of debility, they are alike adapted; and a gradual return to health and vigor will generally be the happy reward of steady perseverance in the employment of our medicines for a shorter or longer period, according to all the peculiar circumstances of the case.

In those cases where nocturnal emissions are of such frequent recurrence as seriously to menace the constitution of the sufferer, the first great object of treatment should be to check this unnatural and most injurious drain upon the system. This effected, we have then to repair the previous mischief, by removing all morbid irritability; giving tone and power of retention to the seminal vessels, the relaxed condition of which is the very foundation of all the mischief.

The very essence of successful treatment consists in first arresting the unnatural loss of that fluid, on the healthy retention of which depends so greatly the vigor of both mind and body. The rest can readily be accomplished by perseverance, and regularity in taking our medicines and attending to the general instructions.

Some writers strenuously advocate the employment of Galvanism, or Electricity, as a remedial agent for the cure of Spermatorrhœa; now, although we have ever been most prompt to test impartially, every so-called remedy which appeared to afford a reasonable prospect of success or assistance, and although this among many others has been most carefully, regularly, efficiently, and perseveringly applied by us personally, yet (and with regret we say it) we are bound to confess that it has fallen far, very far, short of the high encomiums passed upon it by certain of its advocates.

Still, as a remedial agent in many other complaints, it ranks undeniably and deservedly high; as, for example, in several of the varieties of paralysis, partial, local, or general; loss of voice, chorea, or St. Vitus' Dance, epilepsy, apoplexy, etc. In all these and many more affections, galvanism properly applied by, or under the direction of, a medical man (and in other hands infinite harm instead of good will be quite as likely to ensue), possesses remarkably curative powers.

Many other remedies (real or imagined) and

modes of treatment have from time to time been
put forth, and advocated with all the energy
and tenacity of actual conviction as to their in-
trinsic value, but with very few exceptions they
have speedily sunk into merited oblivion. Almost every writer on these matters, past or
present, lends his advocacy to some novel or particular line of treatment, as though he really had
faith only in the theory of his own adoption, to
the exclusion of all and everything besides. In
one voluminous work now before me, published in
the early part of the present century, the author
in some five hundred pages of letter-press,
labors to promulgate his very decided and un-
qualified opinion that Cantharides (Spanish
Fly!) is the only remedy for every ailment of
the generative organs ; and yet this man was of
no mean reputation in his day. In another work
of somewhat more humble pretension, blisters
are as strenuously recommended. Another
advocates setons in the perineum ; this is even
worse than the plan now almost exploded, of
catheterizing the urethra with the porte caustique
armed with nitrate of silver, as introduced some
years ago, by the late gifted and skilful Lalle-
mand of Montpelier. Others, again, advocate
the injection of astringent lotions into the ure-

thra, but we have never had the good fortune to meet with an instance of success, though many cases of failure have come under our notice.

There are some even among living practitioners, who profess to have faith only in preparations of steel or iron, assisted by alteratives, cold bathing, etc., and, indeed, this is the old-fashioned stereotyped routine practice; in reference to which be it said, that when the seminal ducts are so far relaxed that they allow the escape of semen with the urine, and on going to stool, and when the losses occur with unusual frequency during sleep, or by day, and the blood has become thereby seriously deteriorated, some good may possibly result to the bodily health from the employment of ferruginous medicines, but a relapse of the local weakness soon shows any apparent benefit in this particular not to be lasting. Another very important circumstance in reference to the employment of iron, is its known tendency to excite the circulation, and unduly accelerate the action of the heart, shown by the distressing palpitations, headaches, flushings, etc., which often occur during its use. Hence it is contra-indicated in heart and brain diseases, or where there exists any tendency to these affections, which the use of steel or iron would be likely so seriously to aggravate, as to cause sudden death.

The Nitrate of Potash was largely employed and recommended by the late McDougal, but as several of his former patients have, from time to time, come under our care, suffering from the relapse of the symptoms for which he thus treated them, we simply record the fact as evi-dence of its untrustworthiness as a remedy.

The Chlorate of Potash in its action on the system closely resembles the preparation im-mediately preceding, and is therefore equally disentitled to credit as a curative in these com-plaints.

Ergot of rye, camphor, hyoscyamus, opium, digitalis, lupuline from the strobiles of the hop plant; nux vomica, or its alkaloid strychnine; copaiba, cubebs, Indian hemp, etc., have each their advocates, and are all good in their way, though only as adjuncts to other treatment, as they are powerless to effect a permanent cure.

Strapping or compressing the testicles is use-ful in Varicocele, and in such cases is of positive service in reducing the calibre of the vessels and diminishing the supply of blood to the part; but as recommended by some as a remedy for

Spermatorrhœa, it may be classed with the dry cupping; embrocation to the spine and perineum; the suppositories to place in the rectum (a very novel, very useless, and very nasty idea, to say the least) recommended by others, and which are all just so many harmless placebos, concocted solely for the amusement of the patient, by whom they may be either used or left alone with equal benefit. This, however, cannot be said of the mineral acids, as uselessly employed by some men, too often to the serious and permanent injury of their patients' teeth.

We have placed before our readers the usual remedies adopted for the treatment of Nervous and Physical Debility and Impotence. Unfortunately it too often happens that the deluded victim of Self-Indulgence, after endeavoring in vain to find relief from the consequence of early error—sinks at last into a state of lethargy.

Now, in concluding this part of our subject, we may be allowed to observe that it is right and useful that all men should know that there are principles of personal management which cannot be violated without the incurral of grievous penalties; it is right they should know, when wisdom and regret succeed the heyday of inconsiderate self-indulgence, how these penalties may be mitigated, and how the sting of their remorse may ultimately be removed. The most absurd of all emotions is that of despair. Let the sufferer remember that there is scarcely any degree of weakness or functional derangement to which the timely aid of science cannot apply a cure.

CHAPTER V.

ON MARRIAGE; ITS OBLIGATIONS AND EXCESSES; WITH A FEW REMARKS ON MATRIMONIAL UNHAPPINESS.

Marriage is an institution of divine origin. It is the bond of union between the sterner order of our nature, and the gentler sympathies of the female. When a beloved object is about to be united to her faithful suitor, the heart is fraught with the most ecstatic feelings. The various disappointments that may have beset our path are all about to be dispersed; the difficulties that arose in our way are now no more, and the blushing object of our regard has consented to repose within our embracing and endearing arms.

We are not only about to fulfil a natural, but a moral and divine injunction. God has

commanded us " to increase and multiply " our
species ; we have been desired to leave the early
home of our youth ; the tender solicitude of our
parents, and all the-sweet communion of the
hearth circle to which we have been accustomed
from our happy childhood, and to fulfil our
promises of fidelity to our well-chosen spouse,
to cleave unto her, that our union may be
" twain in one flesh."

To persons properly constituted, mentally
and bodily, there can be no greater happiness
than that derived from the mutual intercourse,
the mutual love and endearments of an affectionate couple bound to each other in the lawful
bonds of matrimony. The " Cynthia of the
minute " has no charms comparable to the connubial delight of a fond indulgent pair. The
frail one has nothing to bestow but the vehicle
of sensuality, the possession of which " filthy
lucre " can obtain at any time. Her charms are
common property ; her blandishments are un-
real; her smile a hollow mockery of affection.
The caresses she bestows on the ardent youth
are transferred to tottering imbecility and age ;
while the young wife, " lovely as she is good,
and good as fair," has in the full plenitude of
her power surrendered herself into the em-

braces of her "one true husband." Marriage, however, is not altogether made up of "sighs and wreathed smiles." Though it has its devotions, it has also its obligations; and the divine command, "increase and multiply," can only be obeyed by those in the full possession of mental and bodily vigor; by those who have preserved the golden stream until the time of its flood; who have not plucked the fruit until the day of its juicy ripeness. To such happy creatures the nuptial bed is indeed redolent with entrancing joys. The cares of life are swallowed up in the ample provision that bountiful nature has made for her devoted servants. Enjoyment is in his power, and his arm need only be stretched forth to obtain and to possess. Yet we are not permitted to be lavish of our possessions. The tree must not be stripped of all its fruit; or we must tarry till the spring shall renew the fallen leaves. Excessive indulgence must not drain the cup of pleasure to its dregs, and then expect that pleasure shall still exist. A prolific cause of much unhappiness among the married, is excessive sexual intercourse; it destroys the life on which it ought to feed; it sows the seeds of misery within the hallowed pale of wedlock.

But what stamps effectually the seal of nature's reprobation on excessive matrimonial indulgence, is its destruction of the health of woman. Is it not a most prolific source of those distressing female complaints which bury half our married females prematurely, and seriously impair the remainder? Do not thousands of our women die in consequence? Many a husband has buried more wives than one, killed outright, ignorantly, yet effectually, by the brutality of this passion. As over-eating diminishes the power of appetite, so excess engenders those diseases which cut off this very pleasure. By causing the prolapsus uteri, albus, etc., it renders this intercourse utterly repugnant—mentally, and painful physically; thus inducing the penalty in the direct line of the transgression. It prevents or impairs the offspring. Whatever enfeebles or diseases the sexual apparatus, of course impairs its products, or else prevents offspring altogether. This indulgence causes barrenness.

It deteriorates woman in the estimation of man. Lust carries with it the feeling and sense of degradation. He who indulges frequently, even with his lawful wife, cannot but associate her in his own mind with this debased feeling to which she administers. He first debases her by his brutality, and then despises her for being debased. It is a law of mind that this excess should produce contempt for its partner. The libertine never speaks well of women as a sex; the reasons are obvious. First, rogues suspect all men of being rogues; liars, of being deceptive; and the sensual of sensuality. The opinion that Pope expresses, when he says that

"Every woman is a rake at heart,"

comes under the category of suspicion. Then, again, the libertine has only, or mainly, been acquainted with women as a sexual thing, and not as a pure, refined and affectionate being. Her sexuality he has particularly noticed, and this vice he detests in himself, and therefore in her.

The grand rule to be observed in the sexual obligations of marriage, is to insure a reciprocity of feeling in the female while we do not attempt a too frequent recurrence of the act. Without a due attention to this simple but natural maxim, the cares of matrimony will be increased, and the object for which sexual intercourse was established, will be entirely frustrated and destroyed.

We hope our readers may not for one moment

imagine that because we have exposed the evils
of matrimonial indulgence, we are any less disposed to be severe on that indiscriminate indulgence of an illegal description ; far from it. We
are too well acquainted with the shocking results of promiscuous indulgence ; to be eaten
up by piecemeal with sores and ulcers, nausea-
ting and loathsome beyond description—to lose
bone and muscle and nerve by inches; and liter-
ally be eaten up alive ; besides being simul-
taneously tortured with agony the most excru-
ciating mortals can endure. We may indeed
offer remedies for the alleviation of these fearful
visitations, but we cannot say one word in pal-
liation or extenuation of the crime which produces such manifold miseries. We are aware
that this curse ceases not with the original of-
fender, but is justly entailed upon the children
of the wicked " unto the third and fourth generations." We behold, indeed, with pity, the
poor, maimed and hobbling object, his limbs
distorted, his joints dislocated and racked with
pain, his life ever tormented with foul and running ulcers ; his mind feeble and his passions
ungovernable. A celebrated physician on one
occasion made this remark to us, while we were
discussing this subject, "a father's licentious-

9

ness is a more prolific cause of scrofula, consumptions and kindred affections, than any other I can name." We have had cases under our treatment that had broken out in the patient after being dormant for two or three successive generations. What a cause then, for matrimonial reflections. To imagine a pure and tender mother, whose illicitly-corresponding husband had contaminated with this foul disease, bringing forth a child of apparent beauty and lusty health, to be "covered with ulcers, and grow unclean and filthy beyond measure."

Joined to this, the victim of promiscuous intercourse, fearful that any should be made acquainted with his constitutional bankruptcy, undertakes to cure himself; and when he has nearly ruined his body, and undermined the health of his innocent wife, he betakes himself to a medical adviser, only to become convinced that had he applied for assistance in time, his constitution would have been saved—the life of his partner prolonged, and his children lived to bless him by a long and happy, and healthy existence.

How fearful it is to contemplate the havoc that illicit intercourse has made in the peace and happiness of families! The man who brings

the offerings of his manhood unimpaired to the
shrine of Hymen, has indeed presented an uncommon gift ; for how many are there who have
lost their manly prowess in the fields of Venus!
who have trod the deceitful and insinuating
labyrinth of self-abuse, and have rendered themselves unworthy the title of man, and unfitted
themselves for the performance of that duty
which nature demands when the link of affection
has been cemented in marriage.

No man can presume to enter on the list of
reproductive qualities when his organs of generation are incompetent to meet the coming
exigencies of the state. Moral causes have
sometimes interfered to prevent sexual congress
being duly accomplished ; but physical impotence acts more frequently, and decidedly more
influentially in every case that we have ever
noted. We have already hinted at the causes
which lead to impotence ; they are numerous
and somewhat conflicting. The principal are
early incontinence, the debasing habit of self-
abuse, and the rigorous observance of chastity.
The organs of generation, like every other organ
in the human frame, require to be exercised in
order to keep them in healthy condition ; but

this exercise must not be construed into an absolute abuse.

There is an old proverb, that extremes meet,
and it is an absolute fact that the extreme of
indulgence and the extreme of chastity equally
lead to the same result.   Moderation is a virtue,
though total abstinence may not be a crime.
Temperance in all things is the happy medium we
have to observe, as "extremes in good" are said
to "lead to evil."   The anchorites, who, retired
from the companionship of mankind, lived alone
from communion with woman; wrapping themselves in the cloak and covering of mysterious
contemplation, practicing the most rigorous
penance, and observing the strictest chastity,
became so impotent that the organs of generation were, as it would appear, actually dried and
shriveled away, and showed nothing more of
men than if they had never attained the age of
puberty.

It is not to be imagined that because the dis-
positions of the young and beautiful of females
are gentle and affectionate, they will always rest
passive and content under the knowledge that
the marital engagements cannot be fulfilled.
Though they may be pure and unsullied in the
ways of sexuality, the still small voice of nature

teaches them the secret which marriage is supposed to divulge; and a young female full of
animal vigor, waiting with patient expectancy
for the embrace of connubial love, cannot be
supposed to bear the pangs of bitter disappointment without reflecting on its sole and offending
cause. Did she know that the cause of her
baffled expectations had rendered himself impotent by the filthy habits of his own hands,
would she not loathe and detest him? Would
she not spurn him from her arms with the same
feelings as if he had proclaimed himself the
basest murderer? To such imbecile sufferers
we particularly address ourselves. We call upon
them to "hope against hope," to try our helping
hand; and though he may have contracted
marriage in his impotent state, let him confidently rely on our hermetically sealed silence
and secrecy, and he may shortly be enabled to
hold "her whom his soul loveth" in a warm embrace, and complete the duties of mankind as
becomes a competent and vigorous man. Let
us caution him, should he be still disengaged by
marriage, not to brave the risk of embittering
his home and all its social and domestic joys;
not to trust to his supposed abilities when his
course of self-indulgence whispers serious doubts

of his efficiency; let him not indulge the delusive hope that the fair form and glorious majesty
of virtue will be incentives to his weakened and
unhealthy functions; for we tell him that a single
failure in this particular office has frequently
blasted the buds of female affection, and have
sent the impotent pretender in utter confusion
to hide his shame-covered head in some desolate
cave of the earth, "where he passed his days
and died at last, unwept, unhonored and un-
known." Nature is absolute in all her laws; she
will not be governed at will; her subjects must
attend to her commandments, and not dare to
transgress her wise ordinations; when once
offended she is not so soon appeased; and it is
only by the strictest attention to the rules and
instructions which her high priests (the medical
professors) have laid down, that the afflicted can
ever expect to be restored to the state of former
health and strength which they once enjoyed,
and which they have lost by their consummate
foolishness. Hope should ever stimulate the
sufferer to renewed exertions for the recovery
and reinstatement of his lost fortunes; and in
order to engender and keep alive that hope, it
is absolutely necessary that confidence, or faith,
should be made participator in this anticipation.

But as the watchmaker cannot repair a watch without minutely examining every portion of the complicated mechanism; so the surgeon must be made aware of every peculiarity of disease by means of the patient; every symptom must be minutely detailed; every sensation revealed that affects the unhappy sufferer.

CHAPTER VI.

DISEASES OF THE GENERATIVE ORGANS; GONORRHŒA, ITS SYMPTOMS AND TREATMENT; OBSERVATIONS ON SWELLED TESTICLES, STRICTURE AND GLEET.

The gratification of sexual passion is a natural consequence of our physical conformation; its moderate use is beneficial; its abuse highly injurious; for it is but too well known that from the unrestricted connection of the sexes have arisen certain classes of diseases, which more or less deeply affect the human frame. How or when the venereal diseases were first introduced into Europe, are questions that have repeatedly been asked, but never satisfactorily solved. Some maintain that they are of Eastern origin, others contend that they are inherent to humanity, and have always existed as the opposite of the

sexual pleasure ; just as apoplexy would result
from over-indulgence in eating, or as spontaneous combustion might arise from confirmed
drunkenness.   Certain it is, however, these diseases do exist, whatever may have been their
origin, and that they are poisonous and con-
tagious though not epidemical.   The poisons
generated and transmitted by sexual contact are
of a peculiarly malignant and destructive nature.
Numerous constitutions have been victimized by
this noxious impurity, for dreadful havoc is soon
made on the body when the disease is neglected ;
as soon, therefore, as contagion is imbibed,
proper steps ought to be immediately taken,
before it becomes deeply rooted in the system ;
nor are the effects confined to the first sufferer
only ; they are communicated to the innocent,
and too frequently the virtuous wife has become
sacrificed to the husband's adulterous intercourse
with profligate and abandoned women, and the
unborn infant inherits even in the womb, the
baneful fruits of its father's depravity.   The
various organs and parts of the body have their
distinct functions and some portions of the
frame are more susceptible of particular diseases than others.   All poisons permeate the
body and inflame the stomach.   In common

language, one of them, as gonorrhœa, or clap, limits itself to the surface on which it falls; others, like the syphilis, or lues venerea, derange the entire system of the constitution. The latter disease arises from a morbid poison which produces contamination by contact, and is capable of being infinitely propagated. The matter of gonorrhœa, if applied to the skin, or to any secreting surface, produces local inflammation and a peculiar discharge, mostly without breach of surface; while noxious virus or pox occasions where it falls, an ulcerated, ragged destruction of parts, styled in medical parlance, a chancre.

The peculiar secretion of a chancre is frequently taken up by the absorbent vessels of the living system, and conveyed into the mass of the circulating blood; and in its passage through the glands of the groins, nearest the spot originally infected, these bodies are apt to enlarge, inflame, become intensely painful, to suppurate and burst, forming the complication know by the name of bubo. The original sore may heal, yet from the poisonous qualities being communicated to the general absorbent system, fearful consequences may arise, and the body be infected with the virus, inducing constitutional syphilis, showing itself in thoracic inflammation,

and producing diseases of the skin with painful
enlargement of the bones. We will, however,
in our present remarks confine ourselves to
gonorrhœa, the most common, yet an intensely
painful and troublesome disorder, which is rendered by frequent occurrence one of the greatest social evils that can fall to the lot of our
species.

Gonorrhœa is common to both man and
woman—in the former it generally attacks the
urethra; in the latter the vagina, urethra, clitoris
and nymphæ, are all affected. The first symptoms of gonorrhœa is generally an itching at
the orifice of the urethra, sometimes extending
over the whole glans, and a tingling sensation
is felt, but so slight that it only serves to pro-
voke more frequent erections, and a desire for
sexual intercourse. This is followed by incipient inflammation, and its power of awakening
the copulative propensity is so remarkable, that
women of abandoned character can tell when
they have received new infection, from the presence of that excitement, and those desires to
which they are, in an uninfected state, so usually
unaccustomed. This irritation or itching is
presently exchanged for an uneasy sensation,
and the penis assumes a kind of semi-erect

position or fulness; the lips of the urethra are swelled and pouting, and, if turned outwards, are found to be of an unnatural scarlet hue. The whole urinary canal secretes a quantity of mucus, and is indued with a high degree of sensibility. This increased secretion of mattery fluid arises from various causes; a narrowness of the urinary canal takes place from the thickening and swelling of the mucous membrane which forms its lining; a partial retention of the urine follows, and the parts adjacent are diseased. Irritation goes on in the surrounding organs in proportion to the virulence of the attending symptoms, and the disposition of the body to receive and retain infection; as our readers need not be informed that some persons are more intensely susceptible of inflammation than others.

When this irritation has fairly commenced, the common law of organization is naturally fulfilled, and copious effusion from the inflamed surface supervenes. The acuteness of sensible pain is increased and is referred to the course of the urethra, and a discharge, which is at first transparent, watery, and somewhat whitish fluid, is soon changed, loses its transparency, and is glutinous, ropy, of a stringy consistency; it

next assumes an opaque, whitish, yellowish, or
even of a slightly greenish hue.

It seldom happens that there is any considerable pain experienced until after the appearance
of the discharge, and some men of weak and
feeble powers suffer little or nothing either before or after the discharge. Cases vary greatly
with the virulence or mildness of the poison received. In most instances, a great degree of
soreness is felt at times along the under side of
the penis, accompanied by a sense of fulness,
long before discharge takes place. The glans,
or nut, assumes a transparent cast, which is
chiefly visible near the beginning of the urethra.
The entrance of the urinary canal is often found
to be excoriated, especially if the glutinous
matter that is apt to gather round the mouth of
the orifice is not washed away, as it infects that
part by being suffered to remain and harden
round that place. The first painful symptom
experienced by the patient is scalding. This
characteristic of the disease is greatly augmented by sympathy; for the fear of the sufferer disposes the urethra to convulsive contraction, and the stream of urine is suddenly
obstructed, passing in scattered and unequal
spurtings as it escapes with pain and difficulty

from the diseased and irritated canal. With some this inflammation is not very intense, and merely a slight running, with inconsiderable heat and soreness, is observable; this is the case with persons of cold and insusceptible constitutions. Others are very severely afflicted, inflammation runs high, and that painful affection of the penis is experienced, called chordee. This is caused by the altered length of the penis, which the urethra being obliged to accommodate, is bent or curved downwards with great pain. The agony is intense, if an attempt be made to lift the penis towards the belly; this suffering, however, does not last any length of time, if cold applications be used, though it is apt to return when warm in bed.

There are many other concomitant diseases attendant on gonorrhœa—one is the painful affection named phymosis. A chancre or any immediate irritant, may produce this state of the parts, but generally it is observed as supervening upon gonorrhœa. The prepuce, or foreskin, is inflamed and thickened, and it cannot be drawn back so as to uncover the nut or glans; the discharge is then apt to insinuate itself beneath, and ulcerations are the frequent results. This tightness of the prepuce is productive of bad consequences, particularly in the case of a chancre producing irritation; for the glans being between the orifice of the prepuce and the sore producing the swelling, the pus is confined, and it accumulates round the edge of the glans so as to cause painful ulcers in the inner surface of the foreskin. Great caution ought to be observed in the remedial measures for the removal of this affection; and, though the cure is perfectly surgical, an inexperienced hand might induce mortification by a too rash use of the knife. It will be in all cases absolutely necessary to abate the inflammation previous to dividing the prepuce with the knife. Paraphymosis is that condition of the organs when the prepuce cannot be drawn forward so as to cover the head of the penis, but remains swelled, and grasping the glans in strangulating and painful constriction. This disarrangement is often so great as to impede circulation, threatening mortification, and sloughing the glans.

Sometimes the bladder is affected, in which case it becomes more susceptible of every species of local irritation; the patient, under such circumstances, can with difficulty restrain his urine, the discharge of which is attended with violent pain in the bladder and glans, similar to what is felt in complaints of the stone. Sometimes the kidneys sympathize in the general irritation, but this is not of very frequent occurrence.

Sympathetic buboes, or inflammatory enlargement of the glandular parts of the groin, are apt to accompany the progress of gonorrhœa. The characteristics of venereal or syphilitic bubo and gonorrhœal bubo are nearly similar, but require a decidedly different treatment.

Venereal bubo is almost certain to run on to suppuration and burst; whereas gonorrhœal bubo very rarely under proper medical care becomes converted into an abscess. Errors of judgment are liable to occur in discovering the difference, and it is possible to mistake an enlargement (which, in every respect, is truly syphilitic) for that which is purely, and less injuriously, gonorrhœal.

We have had patients reduced to a deplorable state by lues venerea, from the inexperience or ignorance of practitioners ; and our utmost skill has had to be called into action before we could dissipate the ill effects of the unscientific treatment to which the sufferer had been subjected.

Swelling of the testicles is a common symptom attending gonorrhœa. This is a purely sympathetic affection. and appears first as a soft

fulness of the testicle, which becomes tender
under pressure; hardness follows and accompanied by considerable pain. Sometimes the
spermatic cords are affected, and the veins of
the testicles become varicose. The bowels enter
into the general sympathy, and are frequently
attended by cholic pains; nausea is felt and
vomiting ensues; the organs of digestion are
impaired, and general debility necessarily follows. Swelling of the testicle sometimes entirely
removes the discharge, as the absorbent vessels
have taken up the virulent pus and deposited it
in the seat of the pain; at other times discharge
accompanies the swelling, and ceases with its
restoration to its natural dimensions. If the
constitution be irritable, or if the patient indulge in his usual regimen and exercise during
the first stages of gonorrhœa, this distressing
complaint may be expected. Medicines strongly
purgative, and of a saline or acrimonious description, will induce this affection; but a frequent cause of its production is the incautious
exhibition of irritating and astringent injections
used for the cure of the running.

As we have observed, that as this symptom
of gonorrhœal affection is the most painful, so
it is the most dangerous consequence of the dis-

ease, and the patient that tampers with its cure exposes himself to much suffering, and a train of evils of which he has no conception. Another cause of gonorrhœal inflammation is spasmodic stricture. This disease may arise from many causes, and attack individuals of any age, most commonly the young and those who are of an ardent and plethoric disposition.

Inflammatory stricture may be expected from repeated attacks of the clap, as the lining mucous membrane of the urinary canal becomes thickened and diseased, particularly that portion of it that is seated near the neck of the bladder.

Stricture may creep on after marriage, and produce many serious effects, and the person who is afflicted in this way is not in a proper condition to effect a productive intercourse with his partner; since the seminal liquor, however prolific and healthful it may be, cannot be ejected with sufficient force into the womb, but frequently falls close to the extremity of the penis. The pleasure of the female is not only considerably diminished, but her hopes of having children by such a person can have little foundation while this stricture remains. We have known many instances of men who, after marriage, have consulted us on this subject, and we are happy

to add that their complaints have disappeared,
and a smiling family of fine children now reward the pains they took for the removal of
their incapacity.

Excepting stricture, there is nothing so troublesome and so truly injurious to the constitution,
as a confirmed gleet. This chronic, semi-trans-
parent discharge often remains after the ordinary
symptoms of acute gonorrhœa have abated.
Some authorities have been of opinion that gleet
is of a scrofulous character, an opinion which is
strengthened by the circumstance that cold sea-
water bathing has been found very beneficial as
a remedy; others have thought it to be a result
of debility; but this idea, though it may occasionally be correct, is certainly often erroneous,
for the treatment founded on this supposition
has been found to operate injuriously. This
discharge appears not to have any specific quality, but to vary according to the constitution
of the patient, or to the conditions of the parts
affected; and, therefore, it is beyond the bounds
of possibility to lay down any universal system
of cure; for, as in gonorrhœa, what to one might
prove serviceable will be hurtful to another.
The matter of gleet may not be all purulent
matter, but partly a mixture of discharges from

the secretory organs, and from the vesiculæ seminales, when their ducts are affected.

In the cure of gleet many have had recourse to the use of astringent injections, as also the bougie; we have already cautioned our friends against the inconsiderate exhibition of both these remedies; the first may cause stricture to a frightful extent; and the last, especially in the hands of the untaught or unskilful, may lead to results worse, if possible, than the original disease itself. We may also enumerate inflammatory disease and enlargement of the prostate gland as the followers of severe cases of gonorrhœa. We have already hinted its serious effects while speaking of venereal consequences, as engendering complaints of the bladder, and producing torments that destroy all relish for the enjoyments of the most transient pleasure, making the life of man one continued scene of misery, without the hope of amelioration of pain for one moment. These frightful results of human infirmity have, at all times, our most patient investigation, and we flatter ourselves that we have been the means of bestowing the blessings of relief when the last spark of hope was nearly extinct.

CHAPTER VII.

ON VENEREAL AFFECTIONS, THEIR SYMPTOMS AND TREATMENT ; MERCURIAL PREPARATIONS ; THEIR EVIL RESULTS.

The poison producing syphilis is essentially different from that producing gonorrhœa ; as the former contains a poisonous virus which destroys the substance of the surfaces on which it falls : hence its effects are more certain and more frightfully rapid. The chief difficulty in all syphilitic affections, is to discover whether absorption has taken place, and if so, to what extent ; if none of the virus has been absorbed, the disease may be treated with ease and safety ; whereas, when absorption has followed illicit connection, the case is often attended with difficulty and danger.

We must not only remove the external symptoms, but we must penetrate into the recesses of the system, and root out the lurking poison that courses along " the gates and alleys of our body ;" for, unless we can do this, and effectually, too, we only afford temporary relief, and the patient is not only liable to a relapse, but he is capable of contaminating the lovely and the innocent.

This disease is most commonly developed on the body by means of the local sores of a female, producing inoculation, and consequent similar sores, sometimes singular, but occasionally numerous, and affecting for the most part the external genitals. These sores, or ulcers, are named chancre; it is formed on the male organs on the foreskin and glans or nut of the penis, and has a red or angry appearance ; spreads gradually, and if it be allowed to remain neglected, frequently eats into the male organ, and causes its complete destruction. The lapse of time between the application of the poison and the appearance of the chancre is uncertain; it is thought by some to exceed the period alluded to in describing gonorrhœa, but even this is doubtful.

The first symptom is an irritation, or an itching of the part ; this, if the place affected be the glans, is succeeded by an inflamed pimple, small and watery, which soon displays a rapidly enlarging ulcerated sore. A hollow is seen in the centre, extending to beneath the skin, excessively painful and sensitive ; a blush of dark fiery redness is seen round the ulcer, and the skin becomes unusually thickened and firm. The surface of this sore is yellow, with hard and ragged edges; its outline is irregular and there is a feeling of tension to the touch. The genitals are not only affected with these ulcers, but they may be found on any other part of the body, should that particular part be invested with a mucous membrane, as, for example, the lips or nostrils, and sometimes the chancre appears on the frænum, and, in this case, the part is often destroyed; or ulceration passes through it and contributes much to retard the cure. The irritation caused by chancre on the glans occasionally produces a sympathetic affection of the urethra or urinary canal, which exhibits itself in what is generally termed venereal gonorrhœa.

The testicles and scrotum also sympathize and are often much diseased. If a chancre be limited to the external surface it progresses slowly; but if its destructive ravages have extended deeply beneath the skin, mortification will, too frequently, be the result.

There are four remarkable characteristics in venereal chancre; the first is generally known by its circular form, its excavated surface, covered by a layer of tenacious and adherent matter, and its hard, cartilaginous base and margin.

The second form of chancre is unaccompanied by any appearance of induration, but exhibiting an elevated margin, which appears frequently on the outside of the prepuce and seldom exists alone; this is denominated "the superficial chancre with raised edges."

The third form, termed phagedenic, or malignant chancre, is a corroding ulcer without granulations, and distinguished by its circumference being of a livid or red color; and the fourth is a most formidable kind of chancre, called the sloughing ulcer. It makes its appearance by a black spot, which spreads and becomes detached, leaving a deepened and unhealthy looking surface, which is very difficult to remove. This sore is exceedingly painful and encircled with a purple inflammatory circle. Sloughing is an ordinary result of continued or neglected chancre, which frequently ends not only in the destruction of the parts of generation, but, by the encrvating effects on the constitution, in gradual consumption and death.

The venereal poison from the four already quoted kinds of ulcers is usually taken by absorption from the chancre to the glands of the groin; the virus being conveyed along the lymphatics, and in its passage producing inflammation of these vessels. Bubo, or swelling of the groin, is the result of this absorption. It is the consequence of chancre or gonorrhœa. In the case of chancre, that side of the groin is generally affected on which the chancre appears on the penis; in the latter case both glands are at times affected. The real bubo generally begins with a sense of acute pain which proceeds from the tumor affecting the purulent formation. This pain increases, and has a tendency to suppuration, more or less retarded, according to the nature of the constitution. The bubo retains its original position till after suppuration, and then it becomes more widely spread, the suppuration is extensive, the agony considerable, and the skin highly florid. Sometimes scrofulous enlargements of the groin are mistaken for venereal buboes, but they are very different in character, as they are slightly painful and difficult to suppurate.

The venereal disease becomes constitutional by the absorption and transmission of a poisonous virus; first perhaps through a primary ulcer to the groin, and afterwards spreading itself throughout the entire system of the blood-vessels. The circulating fluid being once contaminated, the various solid structures of the body become gradually affected and poisoned.

Buboes in the groin constitute the first degree, then follow pains, cruelly affecting the head, the joints of the shoulders, arms and ankles. There are also scabs and scurf in various parts of the body; sometimes these scabs are dispersed over the body similar to the disease of leprosy.

The symptoms gradually increase, especially the pain, which becomes so intense that the patient is unable to lie in his bed. Afterwards, nodes arise on the skull, shin-bones, and bones of the arms, which, being attended with constant pain and inflammation, at length grow carious and putrid.

Malignant ulcers now seize different parts of the body, but generally begin with the throat, and thence gradually creep by the palate to the cartilage of the nose, which they destroy, and the nose being destitute of its natural support falls down flat. Besides these symptoms the following are observable in a confirmed lues, though they do not appear in all patients, nor at the same time: The skin, especially about the neck and breast, and between the shoulders, is covered with flat spots, like freckles, of a rosy, purple, yellow or livid hue, sometimes distinct, small and round, like lentiles; sometimes more enlarged and extended. They are full of itchy pustules, tetters and ringworms. There are chaps in the palms of the hands and the soles of the feet, with violent itching, from whence proceeds a clear, serous liquor, and the epidermis peels off in large flakes. The skin thus affected abounds with hard, callous, round, pustules, rising a little on the top, generally dry, but sometimes moist, scaly, branny and yellow; frequently on the corners of the lips and the sides of the nostrils, but more especially on the forehead, temples and behind the ears, where they appear in rows, like strings of beads, and gradually creep among the hair. The hair not only falls off from the head, but leaves all parts of the body where it grows. The nails become unequal, thick, wrinkled and rough; afterwards ulcers arise which causes them to fall off.

The inside of the mouth, throat and nose are also affected. The uvula and tonsils become painful, hot, inflamed and ulcerated; pustules appear in the roof of the mouth, which degenerate into round, malignant ulcers, which rot the bone as far as the nostrils. The affection of the throat is a white, slimy looking ulceration, and there is a most offensive discharge, with a fetid breath, the soft palate being not unfrequently completely removed, or hanging in detached portions ; in fact, the upper and back parts of the throat present one vast ulcerated cavity, covered with adhesive matter, and not only does the voice become hoarse, thick, low, but swallowing the softest food is difficult and painful. The lining membrane of the nose is also affected ; the bones and cartilages are affected ; an incrustation forms on the surface, and should this be removed a quantity of bloody mucous matter is seen on the part exposed.

As this horrid disease progresses, it very frequently leaves the face in a loathsome disfigurement; the cavity of the nostrils is exposed from the throat, the natural prominence of the countenance is destroyed, and a disgusting ulceration alone marks the place where the nasal organ once existed.

A syphilitic disease of the bones is the usual consequence of the existence of venereal inflammation of their investing membrane. The bones are affected in various ways ; in the middle exostoses arise, either hard or soft, sometimes with intense pain, sometimes without. The heads of the bones enlarge every way, but

unequally, which produces tumors, pains, difficulty of motion, and stiff joints.   As the caries
increase, they become brittle, and break upon
the least effort.   Sometimes they are so far dis-
solved as to bend like wax.   The long round
bones, as those of the legs, are generally the
first that suffer an attack ; hence, those enlargements on the shins, well known as Venereal
Nodes, which are, in reality, inflammatory enlargement and thickening of periosteum, which
covers them, and passes into the actual disorganization of the bone itself.

A very considerable time after the chancre
has healed the patient complains, in the evening
of each day, of increasing pains and aching in
the legs, or in some particular place on one of
them.   There is not much swelling at first, and
what there is generally disappears toward morning.   Great sensibility of pain occurs at evening again, and the sleep is broken from the
fever and irritation which accompany his rest.

Neither do the eyes and ears escape the fury
of this disease, for the latter are externally
affected with pains, redness and continual itching ; and, internally, are loaded with humors,
the sight is destroyed, and sometimes a suppuration supervenes.   The ears are also affected

with a singing noise, dulness of hearing, deaf-
ness and pain, whilst their internal surface is
exulcerated and rendered carious. While we
are treating of the eyes and ears being so sensibly affected by the syphilitic influence, we
may here observe should the matter secreted in
the urethra, during the progress of clap, be applied incautiously to the eye with a towel in
washing or by the finger, the slightest particle
of that virulent poison is sufficient of itself to
inflame that tender organ to so serious a-degree,
that, if proper remedies be neglected, a total
loss of sight will inevitably be the result.

The treatment of chancre varies very much,
but it is generally both internal and external;
one important point always to be kept in view
is to abridge the duration of the disease, so that
it may not make any serious inroads on the
constitution.

The simplest method is the extirpating of the
chancre, which may easily be effected at its first
appearance, either by incision or caustic—the
latter being the safest mode and that giving
comparatively little pain. But then this method,
safe as it is, becomes almost impracticable, when
the surrounding parts are contaminated, in
consequence of the difficulty in entirely remov-

ing the diseased places; and hence it is most
essential that measures for cure should be
resorted to in the earliest stage of the disease.

The physicians of the "old school" considered
that preparations of mercury constituted an
unfailing remedy for every form of the venereal
disease. "The safest and most commodious
method of Salivation," says Dr. Brookes, "is
by mercuris dulcis, six times sublimed, given
inwardly in the milder pox; or by mercurial
unction, when the disease has got into the
bones." He goes on to describe the effects
which may be supposed to result from this system of treatment. "Fifteen grains of mercuris
dulcis may be given every morning for four or
five days, when we usually observe the fauces
to inflame, the inside of the cheeks to be tumid,
or high and thick, and ready to fall within the
teeth, upon shutting the mouth; the tongue
looks white and foul; the gums stand out, the
breath stinks, and the whole inside of the mouth
appears shining, as if parboiled and lying in
furrows. The inside of the mouth begins to be
whealed, and you may expect to see it in a state
of ulceration, especially about the salivary
glands." We think that our patients would
scarcely submit to this mode of cure, and con-

sent to have their " mouths parboiled and lying
in furrows " by taking mercuris dulcis. And
yet the present age is not free from men who
resort at all times to this dangerous specific.
The celebrated Dr. Hunter was a great advocate
for the mercurial system of cure. Doubtless
the potent quality of mercury was well understood by him, and in his hands may have proved
advantageous in many cases, but there is no
question that many victims have been made to
the practice of administering this powerful
agent, spontaneously in all cases of venereal
contamination.

For we are told by the advocates of the use
of mercury that "should the salivation be
attended with a cardialgia, or violent pains and
torture, at the stomach, perpetual and incessant
retchings, deliquium, cold sweats—there is
great danger to be apprehended." Many forms
of venereal sore are rendered irritable, and evidently disposed to slough and mortify under
the action of mercury, and there are many of
the older practitioners who can recollect the
period when mercury, in the ordinary doses,
failed to act as a remedy, that it was the practice to increase the dose, supposing that a more
complete saturation of the system could alone

arrest the rapid decay this preparation was itself causing.

There is no subject in the entire range of medical and surgical science that demands a greater amount of practical discrimination and skill than to determine when, and to what extent, and under what peculiar circumstances, this powerful specific may be safely used. That its improper and incautious administration has been productive of horrible consequences cannot for a moment be disputed. The young and beautiful have become self-sacrifices to their inexperience, and have swallowed doses of this mineral poison, which has shown its potency in unseemly blotches on their bodies, and horrible scars on their faces.

This dreadful and insidious disease exhibits a most important feature, which is the actual transmission of syphilitic contamination from the parent to the child. Infants may be affected with syphilis in a hundred different ways. Disease may originate in the fœtus, or before birth, in consequence of the impurity of one or both parents. The celebrated Dr. Burns, Professor of Midwifery in the University of Glasgow, in his work on the " Diseases of Women and Children," observes, " Infection may happen when

neither of the parents has at the time any
venereal swelling or ulceration, and perhaps
many years after a cure has been apparently
effected." " I do not," he says, " pretend here
to explain the theory of syphilis, but content
myself with relating well-established facts!"
Premature labor is not unfrequently the indication of this; the offspring presenting a puny,
feeble, emaciated and wrinkled form. The eyes
are red and inflamed; the cry shrill, husky and
wailing; mattery discharges are emitted from
the eyelids, copper-colored blotches disfigure
the shriveled skin of the genitals and hips; the
nostrils are clogged with an offensive scab-like
discharge; the nails come off, and indeed many
children are brought into the world in a state
of absolute rottenness, unfit to bear the atmos·
phere of this earth.

In thus dwelling on the evil consequences of
sensual indulgence and of the ailments of our
depraved habits, let us not be supposed to
administer, in the least degree, to the morbid
feelings and curiosities of the idle; our design
is to hold up the mirror, as it were, to nature;
to show the horrible deformity of vice, and the
loveliness of virtue and innocence. Our little
book is a guide to manly health, and the causes
11

that prevent it. Let us recapitulate our prem-
ises, and see if we have not fully borne out each
particular head of our subject; and then we shall
prove that we have not lost our own or our readers' time. The physical grandeur of man is in
a state of perfection when he possesses every
physical function in its perfect and original
strength. Every function is capable of increasing our constitutional happiness, when it fulfils
its legitimate design; this exercise is allowed
to a certain extent and then it is natural. But
when this natural exercise is overstretched it
becomes unnatural, and consequently painful.
The gratifications of our appetites in eating and
drinking are decidedly natural, but, if over-
indulged in, they become hurtful, and consequently unnaturally taxed. All indulgences are,
therefore, more or less hurtful in proportion as
they are pursued or restrained. We have shown
the dreadful effects of debasing this physical
grandeur by the indulgence of self-pollution;
we have enumerated some of the interminable
evils resulting from excessive abandonment to
sexual intercourse, promiscuous and even matrimonial; but to give their entire histories would
fill thousands of books a hundred-fold larger
than this. We have shown that health is seriously

Injured, that the body is exhausted by this widely wasting cause; and that not only is the physical greatness of our nature destroyed by our own perversity of inclination, but that our children are made the victims of our vicious indulgences.

CHAPTER VIII.

SELF-DIAGNOSIS; OR, HOW SHALL WE ASCERTAIN UNDER WHAT AFFECTION WE ARE SUFFERING?

In consequence of the frequent inquiries made of us, "How shall I know whether I am suffering from spermatorrhœa? what are the symptoms by which I shall be able to recognize it, or by which it will be accompanied?" we are induced to add a few words on this most important point.

The symptoms are infinitely varied, extremely numerous, and differ greatly in different cases, both in number, nature and degree. It will be well, perhaps, first to put the most prominent of them into a tabular form, and then to introduce one or two illustrative cases.

To render this tabulation more intelligible, the symptoms are divided into local, i. e., affections of the generative organs ; bodily, i.e., affections of the muscular, circulative, nutritive, and respiratory systems; and mental, *i. e.*, affections of the nervous system.

In the first place, as being not only most definite in character, but also as indicative of the disease being more than usually deeply seated and confirmed, the local symptoms may be mentioned. They are as follows:

GENERAL SYMPTOMS.

Pollutions * accompanying expulsion of urine.

Pollutions accompanying defecation.

Erection and emission upon slight excitement, such as the mere presence of females or juxtaposition of their dress, etc.

Emissions under similar circumstances, unaccompanied by erection.

Nocturnal pollutions, with or without erection or consciousness.

Diurnal pollutions.

Spermatic urine.

Contraction of the foreskin.

Spasmodic or dull pains occasionally in the organs.

---

* The terms "pollutions" and "emissions" occur to involuntary escapes of seminal fluid.

Varicocele, or varicose veins in the testicles.

Pimples on shoulder and forehead.

Premature emission during coition.

Priapism, or erections apparently without any exciting cause.

Decrease of sexual desire or enjoyment.

Sanguineous emissions.

Diminution in size of the penis and other organs.

Want or imperfection of erectile power.

CLIMAX—IMPOTENCE.

In reference to general symptoms, it is necessary to observe that many, if not all, of these symptoms may occur in and denote forms of ordinary disease; but if produced by spermatorrhœa, they will be aggravated in degree, and will not yield to treatment known to be eradicative of them in ordinary cases. This fact could be illustrated in a variety of instances, but one may suffice. In an otherwise healthy person an attack of indigestion, originating in inattention to diet, will yield to gentle purgatives, tonics, and other well-known means; but if the symptoms of indigestion exist in consequence of the impairment of the nutritive functions by seminal losses, the ordinary remedies

for such symptoms fail to produce their usual
effect, as until the primary cause of the symptoms be removed, the effect will not only continue but increase.   In like manner disorders in
respiration and circulation may arise indiffer-
ently from spermatorrhœa, or from other causes;
in the latter case the remedies usually indicated
for such symptoms will remove them, but not so
if they be caused by spermatorrhœa; and it may
be mentioned that it has been clearly ascertained
that there is no single function of the animal
economy but may not become deranged by long
continued seminal losses.

GENERAL SYMPTOMS—BODILY.

Muscular, Respiratory, Circulative, and Nutri-
tive Systems.

Increased appetite or voracity (in early stages).
Gnawing, and heat of epigastrium.
Uneasiness, sinking, or faintness before taking
    food, followed by disgust or nausea afterwards.
Want of appetite for plain kinds of food.
Weight of epigastrium.
Quickened pulse.
Flushed face.
Acrid eructations.

Acrid heat at the upper part of œsophagus.
Alteration in secretions of liver and pancreas.
Evolution of flatus.
Colic.
Griping.
Difficulties of breathing, and cough.
Distension of stomach and intestines.
Muscular flaccidity.
Excessive mucous secretions.
Irregular action of the heart.
Apoplexy.
Liquid and unnatural stools.
Diarrhœa.
Inflammation of rectum.
Constipation.
Loss of substance.
Cadaverous appearance of skin.
Hollow, sunken eyes.
Extreme sensibility to cold.
Rheumatism.
Loss of hair.
Pulmonary catarrh.
Indolence, or indisposition to exercise.
Lassitude.
Fatigue on slight exertion.

CLIMAX—CONFIRMED DEBILITY.

GENERAL SYMPTOMS—MENTAL.

Nervous System

Restlessness.
Sighing.
Sensation of congestion.
Want of energy.
Uncertainty of tone of voice.
Nervous asthma.
Vertigo.
Want of purpose.
Dimness of sight.
Weakness of hearing.
Aversion to society.
Blushing.
Want of confidence.
Avoidance of conversation.
Desire for solitude.
Listlessness, and inability to fix the attention.
Cowardice.
Depression of spirits.
Giddiness.
Loss of memory.
Excitability of temper.
Moroseness.
Want of fixity of attention.
Disposition to ruminate.

Trembling of the hands.
Sudden pallor.
Lachrymosity.
Tremor from slight cause.
Pains in the back of the head or spine.
Pain over the eyes.
Disturbed and unrefreshed sleep.
Strange and lascivious dreams.
Hypochondrias.

CLIMAX—INSANITY.

CHAPTER IX.

NOTES FROM OUR CASE BOOK.

We will conclude this essay with the narration of a few cases which have occurred in our practice; but, in so doing, we wish it to be distinctly understood that no danger of publicity is hazarded by those who have already consulted us, or who may hereafter place themselves under our care. They are simply selected from hundreds as illustrative of our treatment, or from containing some points of interest, and are published with the full consent and approval of the gentlemen to whom they refer.

CASES.

A. B., a farmer, aged 28. Whilst at school he had been led into the habit of self-pollution, which he had continued for some years. Having become aware of the sin and danger of the act, he had discontinued it; but on attempting to gratify his passions by the natural means, he found to his dismay that he was totally unable to accomplish the sexual act. The penis was incapable of firm and vigorous erection, and a discharge took place before an entrance could be effected into the female organ. His general health seemed good, and his frame robust; this we attributed to his pursuits, which obliged him to be many hours daily in the open air. We would not undertake the case until he had given a solemn promise never again to indulge in the vile practice of Onanism. He was under treatment for ten weeks, during which he had the pleasure of observing a rapid improvement, and he was discharged thoroughly cured.

The following gratifying testimonial was lately received:

"*New Orleans, Jan. 25.*

" MY DEAR SIRS :—I am happy to inform you

that your treatment has been quite successful, although when I first applied to you I scarcely expected any great benefit; for, as I wrote to you at the time, I had been suffering some years from debility and nervousness, which my friends believed to be constitutional; but which I know in my own mind arose from self-pollution. I heard of you from a gentleman who had been under your care, and was induced by him to consult you. I am grateful to say I had not been under your hands one month before I was very considerably better, and in three months I was as well as ever I was in my life. I am, gentlemen,

"Yours ever faithfully and sincerely,

" CH. B."

Drs. LaGrange & Jordan, '1625 Filbert St., Phila.

A gentleman from Boston called upon us about three years ago. He had been married seven years, and was very unhappy in not having children; he possessed extensive landed property, which would leave the family, if his wife should die childless. His age was 33; his wife's 26. We found that, although he had led a free life, his virile powers were undiminished, and that he was in the habit of having a perfect con-

nection several times a week. On examining
his urine by means of the microscope, we detected the spermatozoa, which proved a waste
of semen, and he admitted, though not without
reluctance, that in his youth he had practiced
Onanism. By treating accordingly, we were
successful even beyond our hopes. His lady
became enciente four months afterwards, and he
has now two children.

C. D., captain of an American ship, aged 35;
about ten days ago had acquired a gonorrhœa;
he drank freely the day before. We found him
suffering severely from excessive pain and difficulty in making water; the foreskin had
swollen, forming a " phymosis;" there was a
copious discharge of yellowish matter. C. D.
was anxious to be cured as quickly as possible,
as he was going to the East Indies in a few days.
Ordered aperient and sedative medicine, full
doses of the Napolitaine Pills, and dismissed
him thoroughly well on the eighth day.

Capt. E. H., Baltimore, three months before
consulting us, had observed a small pimple under
the prepuce or foreskin, after a casual connection.
Having rubbed it off, it was succeeded by an

other, rather larger; when, thinking it was the "heat of the body," he took mild aperient medicines. It remained stationary for some weeks, and then began to increase rapidly; he also noticed a large swelling in each groin. Becoming alarmed, he determined, by the advice of a friend, a brother captain, to consult us; but, unfortunately, urgent business detained him some days longer in Baltimore. We immediately saw that the case was very serious; the poison had acquired terrible virulence by remaining so long in the body, and, in spite of the most powerful remedies, he continued to get worse for some days. The chancres progressed so rapidly as to threaten the entire loss of the penis, and both buboes burst, leaving a large cavity in each groin. However, by the most energetic means, he was brought under the influence of the medicine, and began to improve, though at first very slowly. This case required three months' constant attention before a cure was effected, and even then it was necessary to prescribe strengthening medicines for some weeks longer. We instance this case to prove the importance of an early application in case of syphilis. Had this patient applied to us in the first instance, one week's attention would have been quite sufficient to ensure a thorough cure, without pain or annoyance. We were, however, compelled to order him to give up a voyage—fortunately a short one—but, being part owner of the ship, it was of less importance. We have lately received a letter from Captain E. II., stating that he is now thoroughly well, and as strong as ever.

A gentleman, aged 26, consulted us for an inveterate gleet. About three years previously he had contracted a severe gonorrhœa, for which he applied to his family doctor, who gave the usual remedies. But as soon as he discontinued taking medicine, the discharge came on again, though without pain or inflammation. He applied to several medical men in succession, but with the same result. On discontinuing medicine, the discharge reappeared, especially after connection. He felt most anxious to be cured, as he was under engagement of marriage. We are happy to state that in a few weeks after seeing us he was completely cured. He has since married, and there has been no recurrence of the disease.

A gentleman, living in the country, sent some

urine for examination, and stated that he was
suffering severely from gleet, which had rendered
him very weak. The microscope proved the
presence of numerous spermatozoa. We therefore requested him to forward a single drop of
the discharge, packed between two glass slides.
This he did, and we found unmistakable evidence
of the nature of his complaint. We wrote, in
answer, requesting a personal interview. A
short time afterwards he called upon us, and, in
reply to our questions, admitted that he was
subject to frequent nocturnal pollutions; he
also endured great agony from piles, for which
he had been under treatment several times, and
taken enormous quantities of medicine. As it
would have been inconvenient for him to come
often to the city, we arranged to send him the
required remedies by rail, and six weeks afterwards received the following gratifying letter:

DEAR DOCTORS:—According to promise I again
write to you, but really I do not see any necessity for taking any more physic. But I leave
myself entirely in your hands, and, if you think
it necessary, forward a bottle or two to the
same address. There has not been the slightest discharge from the penis for the last four

weeks, nor have I been troubled with nocturnal pollutions. Nothing remains whatever of the piles; and altogether I feel better than I have been for years. Permit me to thank you most sincerely for your kind attention and skilful treatment. I shall not fail to call upon you, to express my thanks in person, when I am next in Philadelphia.

> I remain,
>> Yours, most respectfully,

———— ————

> *Richmond, Va.*

DRS. LAGRANGE & JORDAN:

DEAR SIRS:—Please forward me two more boxes of pills. I have no doubt they will be sufficient to complete the cure, judging from the extraordinary effect of the box I have taken. The discharge is almost stopped, and I no longer have any pain on making water. Address Post-office (till called for).

> I am,   D. H.

> *St. Joseph, Mo.*

DRS. LAGRANGE & JORDAN:—I am suffering from spots and blotches on the face and body, caused, I believe, by syphilis some years ago, although I was salivated at the time, and thought myself perfectly well. Do you think you can do me any good? If so, I will come over to see you. I may say my age is about 35. I am unmarried. My occupation requires me to be a great deal in the open air; in fact, I am a farmer. Enclosed is your fee. Please answer by return, and tell me candidly what you think of my case.

Yours, respectfully,

M. N.

We wrote, requesting a personal interview, and, accordingly, a few days afterwards, M. N. introduced himself. The spots and blotches on his face and body exhibited the true syphilitic character, and were exceedingly disfiguring in themselves, without taking after consequences into consideration. M. N. became our patient, and followed our advice carefully. In six weeks he again wrote, stating that he was perfectly cured.

W. H. K. called upon us early in February, 18—. He complained of greatly impaired health; want of energy, both of mind and body; bad memory; a dizziness in his eyes after reading

12

and writing; a singing and dull noises in his
ears; and inclination to melancholy and low-
ness of spirits, which all endeavors could not
overcome. When in company and addressed
by any one, he had a suffusion of blushes, and
felt exceedingly irritable, which, after passing
away, left a trembling in the limbs, and a sensation as if cold water were running down his
back. Age, 27. Occupation sedentary. Came
from Charleston for the express purpose of consulting us, having heard our names casually
mentioned by a gentlemen. On investigation
we found that he had been practicing self-pollution for upwards of nine years, but had given it
up about four months. After obtaining his
promise to abandon this vile practice, we commenced treating him by the most energetic
means, and after eighteen weeks received the
following note:

DOCTORS: I am happy to inform you that I
feel restored in every respect, and I do not
think I shall require any more medicine. Even
that symptom—frequent blushing—which caused
you so much trouble, and me so much annoyance,
has completely yielded. Acting on your advice,
I am now engaged to be married; and the first

time I come to your city, I shall do myself the pleasure of paying my respects to you in person.

I am, gentlemen,

Very sincerely yours,

W. H. K.

To Drs. LaGrange & Jordan.

We were favored with a visit from this gentleman, Jan., 1870. He reported himself in perfect health, and the father of a healthy boy.

The first officer of the American ship B., called upon us early in April, 1869. He was suffering from a severe case of secondary symptoms. He had noticed chancres soon after leaving Liverpool, for which he had taken blue pills, and used sulphate of copper lotion. He was slightly salivated, but the chancres healed, and he was delighted to find himself "quite well." The passage was fortunately a rapid one (indeed one of the most rapid of the season). We say fortunately, for ten days before his arrival, secondary symptoms broke out with the greatest virulence. His throat was severely attacked, spots broke out all over his body, his hair fell off (a common effect of mercury), and his constitution seemed completely shattered.

We also found, on careful examination and analysis of his urine, that he was suffering from spermatorrhœa; in fact, there was a loss of seed every time he passed water. This case required very prompt treatment, and for three or four days the disease continued to make head against the medicine. However, we soon controlled it; and in the course of seven weeks, Mr. B. was perfectly. restored to health and vigor.

The master of a ship from the coast of Africa wrote to us in March, 1869, from Baltimore. His symptoms were urgent; and he also had imprudently taken large doses of mercury. By prompt treatment we were fortunate in curing him very rapidly. As he was unable to come over and see us, the case was conducted entirely by correspondence. The following is a copy of the last letter received from this gentleman. It is dated April 17th, 1869:

"DEAR SIRS:—I was suffering from a most frightful case of African syphilis, and, after persevering with your medicines for a very short time, am happy to say I am quite well. I should like another small case of Purifying Drops, for which I enclose, so that, if any accident should occur, I shall always have the medicine at hand. I write this as a certificate and testimonial of gratitude, and hope it may be useful to you.

"I am, dear Doctors,
"Gratefully yours,
"___ ___."

Captain W. J., of the American ship, E. C., writes from New York, June 13th, 1869:

"Doctors:—Your French Specific cured me of a bad gonorrhœa in three days. Accept my thanks, and send me six cases more for my medicine chest, for which I enclose the money."

Remarks.—In this case, as in the previous one, we had not the pleasure of seeing either of our patients. Our rapid success in curing them was due to the great care with which we elicited every symptom, to the prompt treatment we invariably adopt, and also to the intelligence and regularity with which they followed out our instructions and took our medicines. In the latter case, we were satisfied, by an analytical examination of the urine, that there was present only an uncomplicated, though virulent gonorrhœa, and treated accordingly. Owing to the facilities afforded by the admirable postal regulations, much of our practice is carried on by correspondence; and, from our long experience, we do not often find it necessary to require a personal interview, especially when inconvenient to the patient. At the same time, care is requisite to be as minute as possible in the statement of the symptoms.

Mr. N., provision dealer, from one of the towns in the neighborhood of the city, called to consult us in February, 1868. His age was about 30. He had been under medical treatment for upwards of three years, for what was considered secondary symptoms of syphilis. On inquiry, we found that he had suffered from chancres five years previously, which yielded in a few weeks to treatment, not without some little trouble; a bubo having formed, which had been punctured with a lancet, as a slight remaining scar still testified. He remained perfectly well for nearly two years, when small spots, nearly colorless, and scarcely rising above the skin, made their appearance on the scrotum and surrounding parts, also a few on the arms and legs. At first he scarcely noticed them, merely took a little cooling medicine; but the itching and irritation became so great that he determined on having medical advice.

The medical man who had previously attended him considered that it was an attack of secondaries, prescribed mercury (to slight salivation), hydriodide of potassium, and large quantities of sarsaparilla.

Under this treatment Mr. N. remained for about eight months; the disease apparently made no progress, but, on the other hand, he did not improve, and the paroxysms of irritation were as frequent as ever. Going to New York, he consulted a physician, who advised him to remain for a few days. He first commenced treatment with large doses of bichloride of mercury (corrosive sublimate), alternating with quinine and the mineral acids. Under this plan he improved a little, passed a few quiet nights, and began to have hopes of recovery. But the improvement was not lasting; less than a month after his return to business the disease made its appearance as severely as ever. It would be wearisome to detail the history of the case the two following years. Suffice it to say that he placed himself under hydropathic treatment, that he was twice salivated by different medical men whom he consulted, that he persevered in a course of arsenic (liquor arsenicalis, Fowler's) for nearly six months; and that although he sometimes was comparatively well for a month or six weeks, on the whole the disease made steady progress, and was daily becoming worse.

February 7th, 1868. He complains of sensations of itching, burning and stinging, which nearly make him mad, more particularly in bed at night. He had not slept an hour at a time for several nights. Rubbing the parts (the disease is principally confined to the scrotum and surrounding parts) invariably makes him worse, the relief being momentary, but he can scarcely refrain from tearing his flesh. Withal, his appetite is good, general health much better than could have been anticipated, constitution sound.

On making a careful analysis of the urine, we were convinced, as previously from examination, that there was no trace whatever of syphilitic virus. In fact, that no connection existed between the disease and the chancres from which he had suffered. This he would scarcely believe, but we determined that he should decide by the result. Our diagnosis convinced us that the complaint was a severe form of skin disease— " prurigo," with patches of " psoriasis," diffused

on the body. Mr. N. consented to place himself under treatment, without much hope of
relief.

February 14th, 1868. Mr. N. called again,
according to appointment. The irritation during
the week had been worse than ever; he was
nearly mad, and utterly reckless and despairing.
In fact, it is difficult to imagine the torture of
these cases, aggravated, and almost made unen-
durable from want of natural rest. Withal,
there was a very decided improvement, and the
case was progressing in a very satisfactory
manner. But we refused to continue treatment
unless he would promise most solemnly to remain one month under our care. We continued
medicine as before, and, in addition, gave lotions
for outward application; not with the idea that
they would be useful as remedies, for that was
out of the question, but to endeavor, if possible,
to allay the local irritation.

February 21st. Not much better. Still some
mitigation of the heat; the itching not as severe.
Had slept two hours undisturbed the night
before. Continue medicines; change lotions to
be applied after a warm bath.

March 1st. Decided improvement. Two
nights' comfortable rest. Psoriasis dying away

on the body; burning in the scrotum entirely gone.

March 21st.   From the beginning of the month, in spite of one or two slight relapses, the improvement has been steady. He is now apparently well. But such is the treacherous nature of these complaints, that we advised Mr. N. to continue medicine for at least three months longer.

August 29th.   We met Mr. N. to-day. In answer to our inquiries he said, " No return of my complaint, Doctors, or you would have seen me without delay."

Our principal object in inserting this very interesting case is to prove that we must be careful in discriminating what is due to syphilis from what has really no connection with it. This is not easy, and can only be acquired by experience.

General B. writes:—"After trying hundreds of advertised nostrums, I was induced by a friend of mine to commence a course of your medicine for spermatorrhœa and generative weakness, brought on in the first instance by the unfortunate practice of self-pollution, and much aggravated by sexual excess.

"I am sure you will pardon my peevish skepticism, and almost ungentlemanly sarcasm during our first interview, and attribute it to the true cause—a wretched state of health and unstrung nerves. Never shall I forget the kindness and consideration with which you behaved towards me. I need scarcely add that one month's treatment restored me to health and manhood, and I now feel a bounding, vigorous health of mind and body and such as I never before experienced."

This gentleman is now married, and has had no return of the disease.

A gentleman, a partner in a very extensive firm in Chicago, wrote to us as follows:

"*February 11th, 1868.*

"DEAR SIRS:—Knowing your high reputation and professional standing, I am induced to consult you respecting my unfortunate position. I am one who, through ignorance (I may say fatal ignorance), has acted against the laws of God and nature, and injured myself, I fear irretrievably, by indulging in the odious practice of self-pollution. Would to heaven that I had read your book sooner, or that some mentor had

warned me in time, of the consequences of my
sin. I am now 25; I am a junior partner in one
of the largest houses of this city, the firm of
——. I first commenced the practice of self-
pollution at the age of 16 or 17, and have continued to a very recent period, once or sometimes
twice a week. I now feel a heavy dragging pain
in the left testicle, which hangs rather lower
than the other. The penis seems small and
shriveled, and I frequently have emissions at
night. My water is quite clear and apparently
healthy, I therefore do not think there is any
loss of seed in that manner; but there is sometimes a slimy discharge at stool, especially when
I am bound in my bowels, which is frequently
the case. I find myself very weak, and often
have pains in my back. I am very anxious to
marry, but know by experience that my generative organs are too feeble for coition. If I had
married in my present state I know I should
have made myself wretched for life. Nor is this
all; I fear you must be wearied with this miserable confession; but I think it best to give you
my entire confidence. I find my intellectual
faculties are greatly impaired, my memory is
bad, and my nerves unsteady. I frequently
suffer from headache; I feel drowsy and low

spirited, and my voice is husky, and not so strong nor clear as formerly. I think I have now stated everything, and forward, by Adams' Express, a small bottle containing my urine; though, as I stated before, it seems clear and natural. I forgot to mention that when I pass water it frequently feels hot and inflamed at the end. I enclose your fee, and hope that you will give me hopes that my case is curable.

"I am, Dear Sirs,

"Truly yours, etc.,

"_____."

Drs. LaGrange & Jordan, 1625 Filbert St., Phila.

Extract, dated March 25th, 1871 :

"All the urgent symptoms are much abated, I feel more energy, and my spirits are first-rate. Appetite decidedly improved. During the last three weeks I have had only one nocturnal emission, and that was very slight. Please direct my medicine as before."

Extract, dated April 30th, 1871 :

"I consider myself thoroughly cured, and do not think I require any more medicine, though if you desire it, of course I must comply. I have proposed to a young lady whom I have

long loved; we are to be married in about two
months, but my dear Doctors, I will let you
know the date when all is arranged, as I shall
insist on your being present at the wedding.
You will see me next week."

CASES.

ADDED TO THE PRESENT EDITION OF THIS WORK.

IMPORTANT NOTICE.—Most of the following
cases have occurred in our practice during the
present year. But patients are respectfully informed that every case published, or in any way
alluded to, is with the written consent, and frequently at the request of the patient, and that
the most inviolable secrecy may be faithfully
relied on in all communications. Letters are
destroyed or returned to the patient at the termination of every case.

No extra cases will be added to future editions
of this work, and, for convenience of reference,
we have numbered them.

CASE I.

GENERAL DEBILITY THE RESULT OF SELF-ABUSE.

One evening, in the summer of 1869, we were consulted by a young man of a most preposessing appearance. His whole make spoke of a naturally powerful constitution, and his face, when in health, must have been manly and comely. But his step was languid, his bearing like that of an old man, exhausted and life-weary, and his countenance overcast with gloom and anxiety. We questioned him concerning his symptoms. It was the old sad story. He complained of palpitation, trembling of the limbs, dull pains in the back and loins, dyspepsia, constipation, restless nights, frightful dreams, loss of memory and impaired vision; but more especially of depression of spirits and the vague haunting fear of being incurable.

On our inquiring if he had any idea of the cause of his sufferings, he frankly, though with evident shame, ascribed them to the vice of self-abuse, into which he had been enticed at the boarding-school where he was educated, and which he had only recently entirely abandoned. In addition to the constitutional symptoms already particularized, he was suffering from certain local affections. The generative organs were much reduced in size and exceedingly relaxed. He had been annoyed, he stated, by emissions, sometimes to the extent of two or three times a week. These, however, had gradually ceased, whilst his general health, far from improving, had decidedly retrograded. This led us to suspect that he was suffering from spermatorrhœa, the constant and unconscious oozing away of the semen, in consequence of relaxation of the ducts. To place the matter beyond doubt we appealed to the microscope, and upon careful examination detected in the urine spermatozóa in considerable numbers, most of them, however, as is often the case, in advanced stage of disease, broken and mutilated. When we had completed our examination the patient anxiously inquired if we could hold out any hope of recovery, or even of amelioration?

We replied that his condition was certainly very unpromising, but if he would carefully follow out our directions, and, above all, beware of returning to his former malpractice, he might reasonably look forward to a complete recovery. This, our opinion, seemed somewhat to reassure him, and he now confided to us his greatest trouble. He was devotedly attached to an amiable girl, the daughter of an old friend of his father. He felt certain that she returned his love, and that the match would be highly agreeable to both families. But he knew himself to be utterly incapable of performing the duties of married life.

After assuring him that nothing which our long experience could suggest should be wanting for his restoration, we supplied him with suitable restorative medicines, in doses suited to the exigencies of the case. We also furnished him with a lotion for local use, and gave him directions as to his diet and general mode of life. In particular we sought to enliven his mind, and prevent him from brooding over his

13

disease. In a fortnight he called upon us again, by appointment. There was manifest improvement. A careful microscopic examination showed that the imperceptible escape of the seminal fluid had very nearly ceased. Some of the constitutional symptoms were also abating. His sleep in particular was sounder and more refreshing, and he no longer suffered from constipation and dyspepsia. We directed him to persevere in the treatment laid down, with some slight modification which his increasing strength indicated. At his next visit we had the pleasure of ascertaining and informing him that the seminal escape was entirely checked. This great point being gained, our only remaining duties were to guard against a relapse, and to strengthen the constitution and eradicate all traces of the disease. In these points we were completely successful, and, in fact, within three months of our first interview, he was restored to the full enjoyment of health and perfect manly vigor. He was soon afterwards married to the object of his affections, and is now the rather of a large and healthy family—a circumstance which sufficiently attests how complete was the cure.

REMARKS.—Had this young man contracted marriage in the state in which he first called upon us, the consequences must have been disastrous. For him to have fulfilled the duties of a husband would have been impossible, and the very attempt would have aggravated his symptoms to such an extent as to render a cure highly problematical. The treatment of the case was very much facilitated by the good sense of the patient, who strictly followed out our directions, and never showed the least tendency to relapse into his former errors.

CASE II.

LOCAL DEBILITY.

Soon after the recovery of the patient mentioned in the last section, we were consulted by a young man who made a very rambling, incoherent statement.

He also had been enticed into the habit of self-pollution at a boarding-school, and had practiced it to a greater or less extent for some years.* About his seventeenth year he had abandoned the habit, and had sought to gratify his desires in the natural way. In so doing, however, he experienced very little satisfaction, a highly significant fact. In course of time, being established in business, he deemed himself justified in looking out for a wife. He soon met with a girl who seemed in every way suitable, and accordingly married her.

To his surprise and alarm he found matrimonial intercourse utterly impossible. His repeated efforts were all in vain, and had served merely to exhaust his system. He had been recommended to make free use of stimulants, which proved unavailing. He had consulted a

---

* A very large proportion of our patients declare that they first learned the baneful habit of self-pollution at boarding-schools, and state that it was little short of universal among their companions. This vice, in fact, flourishes wherever boys or men are shut out from the companionship of females. Hence it has been aptly called "The vice of monks obscene."

variety of medical practitioners—some of them of very high reputation, and after taking a variety of medicines and undergoing sundry modes of treatment, was pronounced incurably impotent.

The patient's condition was quite peculiar. His general health was of course far from satisfactory; still there were no alarming and distressing constitutional symptoms. But the generative organs were shrunk and relaxed to an extent such as few practitioners have ever witnessed, and we were by no means surprised at the opinion that this was a case of confirmed impotence. Passive spermatorrhœa was found, on careful examination, to exist, and was doubtless of long standing. But so minute and ill-developed were the spermatozoa that they would have altogether escaped the notice of an inexperienced observer.* In a case of this kind, or-

---

* In such cases a microscope of great defining power is absolutely necessary for the detection and satisfactory recognition of the spermatozoa. In every instance we invariably use the microscope for the detection of disease, spermatorrhœa and loss of semen.

dinary modes of treatment would be of little
avail. We have, however, at length succeeded
in discovering and perfecting a valuable remedy
for impotence, which may be justly termed an
extraordinary and wonderful specific, as it has
peculiar properties for saving and nourishing
the seminal fluid. This we at once ordered,
along with a suitable course of medicine for
strengthening the general system. In a few
weeks'time our patient began to improve rapidly.
We had now to make use of special application
to prevent our patient from putting his newly
restored powers to premature and excessive
use. By this means we succeeded in keeping
him within the bounds of a wholesome restraint
until the cure was perfected.

CASE III.

DEBILITY COMPLICATED WITH PULMONARY DISEASE.

Early in the spring of 1867, a gentleman,
twenty-seven years of age, consulted us by
letter. He complained of profound languor and

weakness, irresolution, confusion of ideas, and aversion to business. At the same time he was suffering from cough, emaciation, night sweats, and pains in the left side and under the shoulder blades. He had been treated for pulmonary consumption, but his medical attendant found something abnormal and puzzling in the case, and having heard of several persons who, under our care, had recovered from very advanced stages of debility, he was induced to apply to us.

As his letter was, on several points, incon-clusive, and as he lived in the immediate neigh-borhood, a personal interview took place. We found that he was indeed laboring under a pul-monary affection, coupled with a general nervous prostration. He owned that he was addicted to self-abuse, a habit which he had never for a moment supposed to be injurious. Indeed, we had great difficulty in convincing him that this unfortunate practice was at the root of his malady. We found that he was troubled with involuntary emissions, both during sleep, and even in the daytime.

We informed him that, if he wished us to undertake the case, he must completely abandon the habit of self-abuse, otherwise there could not be the remotest prospect of his recovery. To this, after some hesitation, he consented. We next undertook the three-fold task of checking the local drain upon the system, arresting the consumptive symptoms, and combatting the nervous weakness. In this difficult matter we were often rather thwarted than assisted by our patient—one of that numerous class who, because they have met with some success in business, believe that they know everything. Sometimes he omitted to take the medicines ordered, and broke through the rules we had laid down for his general conduct. Sometimes he thought proper to take remedies which had been recommended him by acquaintances, and once or twice he even relapsed into his old error of self-abuse. Repeatedly we were on the point of declining the responsibility of treating so intractable a patient. At last, finding that every deviation from our advice was accompanied by a relapse, and that we never failed to detect him, he became more rational, and conformed steadily to our directions. From that time his recovery progressed steadily, and, in ten months from the commencement of treatment, we had the pleasure of pronouncing him perfectly cured. This case, but for the folly and obstinacy of the patient, would have been brought to a satisfactory issue in less than one-half the time.

At our last interview with this gentleman we strongly advised him to marry, lest he might be tempted to revert to his former unnatural and injurious habits. We have since learned that he has followed our advice, and with most satisfactory results.

REMARKS.—Pulmonary consumption—though not a direct consequence of self-abuse—is often indirectly brought on by this unnatural indulgence. Wherever a tendency to this fearful evil exists, any cause which lowers the tone of the system, and acts as a drain upon the vital resources, may call the latent evil into full play.

Now, as nothing can be more debilitating and exhausting than self-abuse and its more immediate results, spermatorrhœa, we need not be surprised at finding the votaries of secret indulgence so often fall a prey to pulmonary consumption. Difficult as is the successful treatment of this disease, under any circumstances, it becomes simply impossible when the strength of the patient is constantly drained away through a channel of which the physician is ignorant, and which he cannot therefore check. We are convinced that many young men who die from consumption might be saved if the evil were only traced to its ultimate source, and attacked there. Indeed, in all cases where the young are seized with gradual progressive decay, we should advise their friends and medical attendants to ascertain whether self-abuse and its immediate results be not in part, at least, the cause.

Case IV.

EXCESSIVE INTERCOURSE.

A young man residing at —— consulted us. He had never been guilty of self-abuse,·but had from an unusually early age been addicted to sexual intercourse to a surprising extent. He had married before reaching the age of twenty, and though he had from that time abstained from promiscuous intercourse, he had indulged with his wife till the health of both became seriously affected. He complained of dizziness, determination of blood to the head, dimness of sight, and a failing memory, together with a rapid decline of strength. Conjugal intercourse, which had for some time been very imperfect and unsatisfactory, became altogether impracticable.

On microscopic examination, we found that he was suffering from passive spermatorrhœa, a conclusion which the constitutional symptoms decidedly corroborated. We ordered him at present to abstain entirely from intercourse, and directed a course of powerful medicines, together with our remedy for impotence, and certain other preparations which the peculiar features of the case indicated. As the young man had naturally a vigorous constitution, and had faithfully carried out our directions, the cure was completed in seven weeks.

As soon as he was satisfied with his own progress, he consulted us on behalf of his wife, who had suffered very much from their mutual excesses. This patient's case presented many difficulties, but we are happy to say that in about three months' time she was completely restored to health. At the conclusion of the case, we seriously admonished these young people, as they valued their own happiness, to guard against a repetition of their former error.

REMARKS.—Though excessive sexual intercourse, conjugal or irregular, is much less hurtful than the unnatural habit of self-abuse, it may, and often does, when persisted in, produce results of a very similar kind. Many persons

foolishly imagine that if married they may indulge to any extent with impunity. This is an
utter mistake. It must further be remembered
that matrimonial excess in many cases prevents
conception, and thus frustrates the prospect of
offspring.

CASE V.

SEDENTARY HABITS AND INTENSE STUDY.

A gentleman, aged thirty-one, consulted us
on account of rapidly increasing prostration of
body and mind. His employment compelled
him to spend a very large part of his time in
the most complex and abstruse mathematical
calculations, and his scanty leisure was devoted
to abstract scientific research. According to
his own statement, he found himself suddenly
attacked with pains in the head, dimness of
sight, forgetfulness, and especially with a confusion of ideas which threatened to render him
totally unfit for his position. He found himself
frequently overcome with sorrow, even to tears,
without any ostensible cause. His sleep was

also restless and unrefreshing, and he labored under confirmed dyspepsia and the most obstinate constipation. He had consulted various physicians and surgeons, without receiving any material benefit.

Here, too, the existence of passive spermatorrhœa was proved by microscopic examination. But whence had it arisen? We very soon satisfied ourselves that our patient had neither been addicted to self-abuse nor to sexual excesses. He stated, and we have every reason to believe with perfect truth, that he had never had any intercourse with a female. It was plain that his sedentary habits had brought on constipation; that intense study had lowered the energies of the whole system; and that these two causes, together with the weakness of the generative organs springing from total disuse, had established spermatorrhœa.

In this case it was first of all needful to overcome the obstinate constipation, which, by occasioning chronic irritation of the rectum, had sympathetically affected the spermatic

ducts ; and then to obviate the mischief already
done, and impart new vigor to the enfeebled
system.   For these purposes we directed, as far
as possible, an entire alteration of the patient's
habits.   We advised him to devote all his leisure
to bodily exercise and to enlivening society.
The constipation yielded to the employment of
a purgative specially adapted to such cases,
whilst the general debility was overcome by a
course of regenerating medicines.   In two
months he was completely restored to health,
and could resume his duties to his entire satisfaction.

REMARKS.—Some persons, and even medical
practitioners, entertain the mistaken notion that
spermatorrhœa can never arise except in consequence of self-abuse or sexual excess.   This
mistaken and very uncharitable view has, to our
certain knowledge, deterred persons who have
been affected with this disease from other
causes, from seeking professional aid. Pernicious
as sexual intercourse when carried to excess
undoubtedly is, its total neglect is, in persons

of ordinary constitution, little less injurious, and may lead to the very same results : a fine illustration of the truth of the old saying, that " extremes meet."

CASE VI.

RESULTS OF INTESTINAL WORMS.

A New York merchant, aged thirty-six, applied to us under symptoms different from those described in the last case. He had been married for upwards of ten years, and had several children. Latterly, however, he had suffered from profuse and oft-repeated nocturnal emissions, irritation of the rectum and bladder, causeless and painful erections, together with dyspepsia, marked and rapid decay of strength, and great mental depression.

The patient's habits were regular, active and temperate ; he had never been guilty either of self-abuse or conjugal excess ; what, then, was the cause ? On a little reflection, some peculiar features of the case led us to suspect the presence of intestinal worms. Further investigation confirming our opinion, we ordered a suitable anthelmintic, which caused the evacuation of a large number of ascarides. Upon their removal the local irritation at once subsided, and the emissions ceased.

A short course of regenerating medicine reinvigorated his constitution, and in four weeks he felt, to use his own words, " like himself again."

REMARKS.—This is an exceedingly instructive case. No disease can cease so long as its cause remains in operation. Not all the strengthening and restorative medicines in the world would have given this patient relief, so long as a constant irritation was kept up by the presence of the worms.

CASE VII.

INVETERATE GLEET.

An engineer had suffered from gleet for five years. He had consulted several surgeons, and, according to their direction, had used a variety of injections and inward remedies. He had

14

sometimes received temporary benefit from treatment, but upon fatigue, exposure to cold and wet, the use of stimulants, or sexual intercourse, the disease never failed to re-appear. The discharge was ordinarily very trifling in amount, but by its long duration it had somewhat impaired his general health. He was moreover engaged to be married, but had been obliged to put off the affair from time to time, fearing that he was still in a state to communicate infection.

Having subjected specimens of the discharge to a careful microscopic examination, in order to ascertain whether the affection was complicated with spermatorrhœa, we procceded at once with the active treatment of the case. We directed the use of the " Remedy," which, in ordinary and recent cases, is alone sufficient to control affections of this nature. But, as the disease had become so inveterate, and the parts had acquired what may be called a morbid habit, we ordered a course of active and strengthening medicine. Regenerating treatment was finally used to invigorate the general system, and in three weeks' time we had the pleasure of declaring him to be radically cured and in a fit state for marriage.

CASE VIII.

AGGRAVATED GONORRHŒA, COMPLICATED WITH SPERMATORRHŒA.

A young gentleman suffering from spermatorrhœa in its active form, in consequence of self-abuse, unfortunately contracted gonorrhœa. By the advice of a friend he had procured an astringent injection, which, however, he had discarded just in time to escape an attack of orchitis, in consequence of its injudicious application. The local swelling and inflammation were intense, the discharge not very plentiful, but dark-colored and occasionally streaked with blood. The nocturnal emissions were frequent and profuse and were attended with great pain of some hours' duration.

Our first object was, of course, to remove the gonorrhœa, which, aggravated as it had been by improper treatment, rapidly yielded to the Specific Pills, accompanied by certain subsidiary remedies, and within ten days was entirely arrested. The nocturnal emissions, however, continued as abundant as ever, though no longer attended with such severe pain. These we also succeeded in preventing by appropriate remedies, and finally we strengthened the constitution of the patient by the regenerative treatment.

REMARKS.—This case, if treated with the ordinary compounds of capivi, cubebs, etc., would infallibly have been attended with serious complications, and in all probability would have left behind it the harassing sequels of gleet and stricture. By our system of treatment the malady was totally eradicated without leaving any lurking mischief behind, and without injury to the constitution. And this result, moreover, was achieved in a patient who had been previously addicted to self-abuse. Now it is well known that specific diseases in such persons are usually obstinate and uncontrollable.

This case strongly illustrates the folly of those who, when attacked with some specific affection, undertake their own treatment. Had this patient consulted us at once, without making use of his injection, he would have been radically cured in two or three days, and with much less both of expense and suffering.

CASE IX.

GLEET AND SPERMATORRHŒA, THEIR SYMPTOMS, ETC.

A young man engaged in a large drapery establishment where many hands were employed, consulted us a few months ago for an obstinate gleet. It was at once evident that he had been a votary of self-abuse, indeed he said he could scarcely escape, as all his companions were more or less addicted to the habit. Some two or three years back he read one of our books, and became thoroughly alarmed and disgusted with the propensity. He at length had recourse to illicit intercourse, and thus contracted gonorrhœa, when, through unskilful treatment, gleet was the sequelæ. In this dilemma he consulted us. Now, about this time his symptoms were complicated and many. He was extremely weak, losing flesh, headache, cold perspirations, eyesight much affected, especially left eye, frequent dizziness, especially when stooping, emissions, pains in the shoulders and spinal column, urine thick, passed frequently in small quantities. The result of our treatment was the complete recovery of this patient, and being restored to health, he sought up all those young men whom he knew to be guilty of Onanism—warned them of their danger, and induced them to apply to us for the necessary treatment. They all had the good sense to follow our instructions, etc., and were cured. Thus, through the candor, conscientiousness, and moral courage of one young man, a number of others were rescued from vice, disease and misery, and brought back to health and happiness.

Reader, do you know any youth who is gradually succumbing to the effects of secret vice? Let such not perish without warning!

Case X.

CONSTITUTIONAL SYPHILIS.

A gentleman, aged thirty-two, consulted us in a state of no little anxiety. He had in youth led a "fast" life, and had repeatedly contracted both gonorrhœa and infection of a more serious nature. For this latter disease he had been treated with mercury in great abundance, and had been pronounced cured. Believing this to be the case, he had married and become the father of two children, both of whom had lived only a very short time. Latterly he had felt his health somewhat declining, and was suffering from eruptions on the head, back, and chest, whilst an ulcer appeared to be forming at the back of the throat. His wife, too, was evidently affected in a very similar manner, with sore throat, copper-colored eruptions, baldness, and very severe nocturnal pains.

The wife, who had not been saturated with mercury, rapidly recovered; but her husband suffered as much from the improper remedies used by his former medical attendants as from the disease itself. Ultimately we succeeded in curing him, both of the infection and the mineral poisons.

CASE XI.

AFFECTION OF THE NERVES AND NEEDLESS DESPONDENCE.

A commercial traveler, twenty-six years of age, consulted us for nervous debility. He had been compelled to relinquish a most satisfactory and lucrative engagement in consequence of loss of memory, bashfulness, and total incapacity for business. Amongst the more prominent symptoms he enumerated were trembling of the hands, pains in the back and loins, and frequent emissions by day and night—the result of a certain unhappy propensity. He never had recourse to sexual connection, being (as he stated) in fear of infection. At the time of his first visit he was quite incapable of following his business, and, to use his own expression, "only felt happy when alone." With this patient we had unusual difficulty. He was sunk in despondence, and, though he agreed to follow our advice, felt persuaded that he was incurable. We soothed and encouraged him as far as possible, and after he had taken our medicines for three weeks we had the pleasure to perceive decided marks of amendment. The emissions had ceased to flow, his sleep was sounder, his digestion better, and pains and tremors disappeared. His mind became also more tranquil, and he felt once more a strong inclination for work. These favorable symptoms increased his confidence; after persevering with the treatment for four more weeks, we had the pleasure of pronouncing him perfectly cured. He is now again on the road, and, as he informs us, can go through his work with pleasure to himself and satisfaction to his employers.

NOTICE TO PATIENTS AND INVALID READERS.

DRS. LAGRANGE & JORDAN, having for many years exclusively devoted their attention to the treatment of the diseases of the GENERATIVE AND NERVOUS SYSTEM described in the preceding pages, may be personally consulted from Ten in the morning till Two, and from Five in the evening till Eight, daily, at their offices, No. 1625 Filbert St., Philadelphia.

We wish our readers to be distinctly informed on the following points in our practice :

1st.—Each patient may always rely on seeing one of us personally on every occasion. We never consign a patient to the treatment of an assistant.

2d.—We are always to be seen during the hours stated, but can be seen at other times by special appointment.

3d.—Every case is strictly confidential.

By strictly adhering to these rules, we have, we are happy to say, been successful in consolidating a practice inferior to none; and have been equally successful in bringing to a happy issue cases which have been given up by some of the first men in the States.

One personal interview, even with patients at a distance, is highly desirable, where practicable. The advantages to the patient are manifold when compared with mere correspondence. A SINGLE VISIT will in most cases enable Drs. LAGRANGE & JORDAN to form an accurate judgment, *and thus facilitate the patient's recovery.* In the first place, many important questions affecting the patient are likely to be suggested by a personal interview, which would in all probability be lost sight of in correspondence. Secondly, a more correct diagnosis of the disorder, and a better appreciation of the patient's constitution, can be arrived at, whilst a microscopic examination of the urine, where necessary, will render any mistake impossible ; especially in cases of spermatorrhœa. And thirdly, where the patient is laboring under urethral discharges, which may or may not be produced by impure connection, one personal visit, with a view to urinary examination, is eminently advantageous ; and the correspondent will be more than repaid the trouble and expense of a journey by the increased certainty and rapidity of cure.

Country patients are informed that they can have the necessary remedies sent to any address or directed to be left till called for at any railway station, in a portable compass, carefully packed and free from observation ; and they

may be taken without confinement or any restraint. It is well that patients should consult
their own interests by being as minute as possible in the detail of their symptoms, age, general
habits of living, occupation, etc. The communication must be accompanied by the usual
consultation fee, $5, or a P. O. order. In all
cases secrecy is considered inviolable, as all
letters are either returned to the writers or destroyed at the termination of each case. When
notes are enclosed in the letters, it will ensure
their safety to register them at the post office.

For a long series of years much of our practice
has been carried on by correspondence only,
distance being no hindrance or additional expense to invalids resident in the most distant
and remote parts of the country. Hence we
have numerous patients whom we have cured
without a single interview, but whose cases have
been conducted entirely in writing.

Patients are requested to keep the same initial throughout the correspondence, and each letter should contain the address to which the writer requires the answer to be directed.

As an examination with the microscope is frequently of the utmost importance, any patient consulting by letter is requested to forward us a flat two-ounce bottle filled with the urine passed on rising in the morning, securely corked and sealed, and packed carefully in wool (to prevent breaking), in a seidlitz box, which, with the flat bottle, may be obtained of any druggist. The parcel to be addressed, CARRIAGE PAID, to Drs. LaGRANGE & JORDAN, 1625 Filbert St., Philadelphia, Pa.

Drs. LaGRANGE & JORDAN particularly wish to impress the propriety, and, in many instances, the absolute necessity, of having, at all events, one personal interview, which, by enabling them to form an exact judgment of the nature of each case, will materially forward the cure.

Drs. LAGRANGE & JORDAN may be personally consulted daily, from Ten in the morning till Two, and from Five till Eight in the evening, Sundays, from Ten till One only, at their offices,

1625 FILBERT STREET,

PHILADELPHIA, PA.

Principal,

R. J. LaGRANGE, M. D.